Wired for Story

WIRED
FOR STORY

The Writer's Guide to Using
Brain Science to Hook Readers
from the Very First Sentence

LISA CRON

TEN SPEED PRESS
Berkeley

Ten Speed Press and the Ten Speed Press colophon are registered trademarks of
Random House, Inc.

Front cover pen photograph copyright © iStockphoto.com/pmphoto
Front cover ink blot illustrations copyright © iStockphoto.com/Krasstin

Library of Congress Cataloging-in-Publication Data

Cron, Lisa.
  Wired for story : the writer's guide to using brain science to hook readers from
the very first sentence / Lisa Cron. — 1st ed.
      p. cm.
  Includes bibliographical references and index.
  Summary: "This guide reveals how writers can take advantage of the brain's
hardwired responses to story to captivate their readers' minds through each plot
element"—Provided by publisher.
  1. English language—Rhetoric. 2. Fiction—Authorship. 3. Creative writing.
I. Title.
  PE1408.C7164 2012
  808.036—dc23
                            2011049478

ISBN 978-1-60774-245-6
eISBN 978-1-60774-246-3

Printed in the United States of America

Design by Betsy Stromberg

18

First Edition

**To my children, Annie and Peter,
the best storytellers I know.**

Contents

9 What Can Go Wrong, Must Go Wrong— and Then Some

COGNITIVE SECRET: The brain uses stories to simulate how we might navigate difficult situations in the future.

STORY SECRET: A story's job is to put the protagonist through tests that, even in her wildest dreams, she doesn't think she can pass.

10 The Road from Setup to Payoff

COGNITIVE SECRET: Since the brain abhors randomness, it's always converting raw data into meaningful patterns, the better to anticipate what might happen next.

STORY SECRET: Readers are always on the lookout for patterns; to your reader, everything is either a setup, a payoff, or the road in between.

11 Meanwhile, Back at the Ranch

COGNITIVE SECRET: The brain summons past memories to evaluate what's happening in the moment in order to make sense of it.

STORY SECRET: Foreshadowing, flashbacks, and subplots must instantly give readers insight into what's happening in the main storyline, even if the meaning shifts as the story unfolds.

12 The Writer's Brain on Story

COGNITIVE SECRET: It takes long-term, conscious effort to hone a skill before the brain assigns it to the cognitive unconscious.

STORY SECRET: There's no writing; there's only rewriting.

Introduction

Once upon a time really smart people were completely convinced the world was flat. Then they learned that it wasn't. But they were still pretty sure the sun revolved around the Earth . . . until that theory went bust, too. For an even longer period of time, smart people have believed story is just a form of entertainment. They've thought that beyond the immense pleasure it bestows—the ephemeral joy and deep sense of satisfaction a good story leaves us with—story itself serves no necessary purpose. Sure, our lives from time immemorial would have been far drabber without it, but we'd have survived just fine.

Wrong again.

Story, as it turns out, was crucial to our evolution—more so than opposable thumbs. Opposable thumbs let us hang on; story told us what to hang on to. Story is what enabled us to imagine what might happen in the future, and so prepare for it—a feat no other species can lay claim to, opposable thumbs or not.[1] Story is what makes us human, not just metaphorically but literally. Recent breakthroughs in neuroscience reveal that our brain is hardwired to respond to story; the pleasure we derive from a tale well told is nature's way of seducing us into paying attention to it.[2]

1

In other words, we're wired to turn to story to teach us the way of the world. So if your eyes glazed over back in high school when your history teacher painstakingly recited the entire succession of German monarchs, beginning with Charles the Fat, Son of Louis the German, who ruled from 881 to 887, who could blame you? Turns out you're only, gloriously, human.

Thus it's no surprise that when given a choice, people prefer fiction to nonfiction—they'd rather read a historical novel than a history book, watch a movie than a dry documentary.[3] It's not because we're lazy sots but because our neural circuitry is designed to crave story. The rush of intoxication a good story triggers doesn't make us closet hedonists—it makes us willing pupils, primed to absorb the myriad lessons each story imparts.[4]

This information is a game changer for writers. Research has helped decode the secret blueprint for story that's hardwired in the reader's brain, thereby lifting the veil on what, specifically, the brain is hungry for in every story it encounters. Even more exciting, it turns out that a powerful story can have a hand in rewiring the reader's brain—helping instill empathy, for instance[5]—which is why writers are, and have always been, among the most powerful people in the world.

Writers can change the way people think simply by giving them a glimpse of life through their characters' eyes. They can transport readers to places they've never been, catapult them into situations they've only dreamed of, and reveal subtle universal truths that just might alter their entire perception of reality. In ways large and small, writers help people make it through the night. And that's not too shabby.

But there's a catch. For a story to captivate a reader, it must continually meet his or her hardwired expectations. This is no doubt what prompted Jorge Luis Borges to note, "Art is fire plus algebra."[6] Let me explain.

Fire is absolutely crucial to writing; it's the very first ingredient of every story. Passion is what drives us to write, filling us with the exhilarating sense that we have something to say, something that will make a difference.

But to write a story capable of instantly engaging readers, passion alone isn't enough. Writers often mistakenly believe that all they need to craft a successful story is the fire—the burning desire, the creative spark, the killer idea that startles you awake in the middle of the night. They dive into their story with gusto, not realizing that every word they write is most likely doomed to failure because they forgot to factor in the second half of the equation: the algebra.

In this, Borges intuitively knew what cognitive psychology and neuroscience has since revealed: there is an implicit framework that must underlie a story in order for that passion, that fire, to ignite the reader's brain. Stories without it go unread; stories with it are capable of knocking the socks off someone who's barefoot.

Why do writers often have trouble embracing the notion that there is more to creating a story than having a good idea and a way with words? Because the ease with which we surrender to the stories we *read* tends to cloud our understanding of stories we *write*. We have an innate belief that we know what makes a good story—after all, we can quickly recognize a bad one. When we do, we scoff and slip the book back onto the shelf. We roll our eyes and walk out of the movie theater. We take a deep breath and pray for Uncle Albert to stop nattering on about his Civil War reenactment. We won't put up with a bad story for three seconds.

We recognize a good story just as quickly. It's something we've been able to do since we were about three, and we've been addicted to stories in one form or another ever since. So if we're hardwired to spot a good story from the very first sentence, how is it possible that we don't know how to *write* one?

Once again, evolutionary history provides the answer. Story originated as a method of bringing us together to share specific information that might be lifesaving. *Hey bud, don't eat those shiny red berries unless you wanna croak like the Neanderthal next door; here's what happened. . . .* Stories were simple, relevant, and not so different from a little thing we like to call gossip. When written language evolved eons later, story was free to expand beyond the local news and immediate concerns of the community. That meant readers—with hardwired expectations in place—had to be drawn to the story on its own merits. While no doubt there were always masterful storytellers, there's a huge difference between sharing a juicy bit of gossip about crazy Cousin Rachel and pounding out the Great American Novel.

Fair enough, but since most aspiring writers love to read, wouldn't all those fabulous books they wolf down give them a first-class lesson in what hooks a reader?

Nope.

Evolution dictates that the first job of any good story is to completely anesthetize the part of our brain that questions how it is creating such a compelling illusion of reality. After all, a good story doesn't *feel* like an illusion. What it feels like is life. Literally. A recent brain-imaging study reported in *Psychological Science* reveals that the regions of the brain that process the sights, sounds, tastes, and movement of real life are activated when we're engrossed in a compelling narrative.[7] That's what accounts for the vivid mental images and the visceral reactions we feel when we can't stop reading, even though it's past midnight and we have to be up at dawn. When a story enthralls us, we are inside of it, feeling what the protagonist feels, experiencing it as if it were indeed happening to us, and the last thing we're focusing on is the mechanics of the thing.

So it's no surprise that we tend to be utterly oblivious to the fact that beneath every captivating story, there is an intricate mesh of interconnected elements holding it together, allowing it to build with seemingly effortless precision. This often fools us into thinking we know exactly what has us hooked—things like beautiful metaphors, authentic-sounding dialogue, an interesting character—when, in fact, despite how engaging those things appear to be in and of themselves, it turns out they're secondary. What has us hooked is something else altogether, something that underlies them, secretly bringing them to life: *story*, as our brain understands it.

It's only by stopping to analyze what we're unconsciously responding to when we read a story—what has *actually* snagged our brain's attention—that we can then write a story that will grab the reader's brain. This is true whether you're writing a literary novel, hard-boiled mystery, or supernatural teen romance. Although readers have their own personal taste when it comes to the type of novel they're drawn to, unless that story meets their hardwired expectations, it stays on the shelf.

To make sure that doesn't happen to your story, this book is organized into twelve chapters, each zeroing in on an aspect of how the brain works, its corresponding revelation about story, and the nuts and bolts of how to actualize it in your work. Each chapter ends with a checklist you can apply to your work at any stage: before you begin writing, at the end of every writing day, at the end of a scene or a chapter, or at 2:00 a.m. when you wake up in a cold sweat, convinced that your story may be the worst thing anyone has written, ever. (It's not; trust me.) Do this, and I guarantee your work will stay on track and have an excellent chance of making people who aren't even related to you want to read it.

The only caveat is that you have to be as honest about your story as you would be about a novel you pick up in a bookstore, or a movie you begin watching with one finger still poised on the remote. The idea is to pinpoint where each trouble spot lies and then remedy it before it spreads like a weed, undermining your entire narrative. It's a lot more fun than it sounds, because there's nothing more exhilarating than watching your work improve until your readers are so engrossed in it that they forget that it's a story at all.

1

HOW TO HOOK THE READER

COGNITIVE SECRET:

We think in story, which allows us to envision the future.

STORY SECRET:

From the very first sentence, the reader must want to know what happens next.

I find that most people know what a story is until they sit down to write one.

—FLANNERY O'CONNOR

IN THE SECOND IT TAKES YOU to read this sentence, your senses are showering you with over 11,000,000 pieces of information. Your conscious mind is capable of registering about forty of them. And when it comes to actually paying attention? On a good day, you can process seven bits of data at a time. On a bad day, five.[1] On one of *those* days? More like minus three.

And yet, you're not only making your way in a complex world just fine, you're preparing to write a story about someone navigating a world of your creation. So how important can any of those other 10,999,960 bits of information really be?

Very, as it turns out—which is why, although we don't register them consciously, our brain is busy noting, analyzing, and deciding whether they're something irrelevant (like the fact that the sky is still blue) or something we need to pay attention to (like the sound of a horn blaring as we meander across the street, lost in thought about the hunky guy who just moved in next door).

What's your brain's criterion for either leaving you in peace to daydream or demanding your immediate and total attention? It's simple. Your brain, along with every other living organism down to the humble amoeba, has one main goal: survival. Your subconscious brain—which neuroscientists refer to as the *adaptive* or *cognitive unconscious*—is a finely tuned instrument, instantly aware of what matters, what doesn't, why, and, hopefully, what you should do about it.[2] It knows you don't have the time to think, "Gee, what's that loud noise? Oh, it's a horn honking; it must be coming from that great big SUV that's barreling straight at me. The driver was probably texting and didn't notice me until it was too late to stop. Maybe I should get out of the—"

Splat.

And so, to keep us from ending up as road kill, our brain devised a method of sifting through and interpreting all that information much, much faster than our slowpoke conscious mind is capable of. Although for most other animals that sort of innate reflex is where evolution called it a day, thus relegating their reactions to what neuroscientists aptly refer to as *zombie systems*, we humans got a little something extra.[3] Our brain developed a way to consciously navigate information so that, provided we have the time, we can decide on our own what to do next.

Story.

Here's how neuroscientist Antonio Damasio sums it up: "The problem of how to make all this wisdom understandable, transmissible, persuasive, enforceable—in a word, of how to make it stick—was faced and a solution found. Storytelling was the solution—storytelling is something brains do, naturally and implicitly. . . . [I]t should be no surprise that it pervades the entire fabric of human societies and cultures."[4]

We think in story. It's hardwired in our brain. It's how we make strategic sense of the otherwise overwhelming world around us. Simply put, the brain constantly seeks meaning from all the input thrown at it, yanks out what's important for our survival on a need-to-know basis, and tells us a story about it, based on what it knows of our past experience with it, how we feel about it, and how it might affect us. Rather than recording everything on a first come, first served basis, our brain casts us as "the protagonist" and then edits our experience with cinema-like precision, creating logical interrelations, mapping connections between memories, ideas, and events for future reference.[5]

Story is the language of experience, whether it's ours, someone else's, or that of fictional characters. Other people's stories are as important as the stories we tell ourselves. Because if all we ever had to go on was our own experience, we wouldn't make it out of onesies.

Now for the really important question—what does all this mean for us writers? It means that we can now decode what the brain (aka the reader) is *really* looking for in every story, beginning with the two key concepts that underlie all the cognitive secrets in this book:

1. Neuroscientists believe the reason our already overloaded brain devotes so much precious time and space to allowing us to get lost in a story is that without stories, we'd be toast. Stories allow us to simulate intense experiences without actually having to live through them. This was a matter of life and death back in the Stone Age, when if you waited for experience to teach you that the rustling in the bushes was actually a lion looking for lunch, you'd end up the main course. It's even more crucial now, because once we mastered the physical world, our brain evolved to tackle something far trickier: the social realm. Story evolved as a way to explore our own mind and the minds of others, as a sort of dress rehearsal for the future.[6] As a result, story helps us survive not only in the life-and-death physical sense but also in a life-well-lived social sense. Renowned cognitive scientist and Harvard professor Steven Pinker explains our need for story this way:

> Fictional narratives supply us with a mental catalogue of the fatal conundrums we might face someday and the outcomes of strategies we could deploy in them. What are the options if I were to suspect that my uncle killed my father, took his position, and married my mother? If my hapless older brother got no respect in the family, are there circumstances that might lead him to betray me? What's the worst that could happen if I were seduced by a client while my wife and daughter were away for the

weekend? What's the worst that could happen if I had an affair to spice up my boring life as the wife of a country doctor? How can I avoid a suicidal confrontation with raiders who want my land today without looking like a coward and thereby ceding it to them tomorrow? The answers are to be found in any bookstore or any video store. The cliché that life imitates art is true because the function of some kinds of art is for life to imitate it.[7]

2. Not only do we crave story, but we have very specific hard-wired expectations for every story we read, even though—and here's the kicker—chances are next to nil that the average reader could tell you what those expectations are. If pressed, she'd be far more likely to refer to the magic of story, that certain *je ne sais quoi* that can't be quantified. And who could blame her? The real answer is rather counterintuitive: our expectations have everything to do with the story's ability to provide information on how we might safely navigate this earthly plane. To that end, we run them through our own very sophisticated subconscious sense of what a story is supposed to do: plunk someone with a clear goal into an increasingly difficult situation they then have to navigate. When a story meets our brain's criteria, we relax and slip into the protagonist's skin, eager to experience what his or her struggle feels like, without having to leave the comfort of home.

All this is incredibly useful for writers because it neatly defines what a story is—and what it's not. In this chapter, that's exactly what we'll examine: the four elements that make up what a story is; what we, as readers, are wired to expect when we dive into the first page of a book and try it on for size; and why even the most lyrical, beautiful writing by itself is as inviting as a big bowl of wax fruit.

So, What *Is* a Story?

Contrary to what many people think, a story is not just something that happens. If that were true, we could all cancel the cable, lug our Barca-loungers onto the front lawn, and be utterly entertained, 24/7, just watching the world go by. It would be idyllic for about ten minutes. Then we'd be climbing the walls, if only there were walls on the front lawn.

A story isn't simply something that happens *to* someone, either. If it were, we'd be utterly enthralled reading a stranger's earnestly rendered, heartfelt journal chronicling every trip she took to the grocery store, ever—and we're not.

A story isn't even something *dramatic* that happens to someone. Would you stay up all night reading about how bloodthirsty Gladiator A chased cutthroat Gladiator B around a dusty old arena for two hundred pages? I'm thinking no.

So what *is* a story? A story is how what happens affects someone who is trying to achieve what turns out to be a difficult goal, and how he or she changes as a result. Breaking it down in the soothingly familiar parlance of the writing world, this translates to

"What happens" is the **plot**.

"Someone" is the **protagonist**.

The "goal" is what's known as the **story question**.

And "how he or she changes" is **what the story itself is actually about**.

As counterintuitive as it may sound, a story is not about the plot or even what happens in it. Stories are about how we, rather than the world around us, change. They grab us only when they allow us to experience how it would feel to *navigate* the plot. Thus story, as we'll see throughout, is an internal journey, not an external one.

All the elements of a story are anchored in this very simple premise, and they work in unison to create what appears to the reader as reality, only sharper, clearer, and far more entertaining, because stories do what our cognitive unconscious does: filter out everything that would distract us from the situation at hand. In fact, stories do it better, because while in real life it's nearly impossible to filter out all the annoying little interruptions—like leaky faucets, dithering bosses, and cranky spouses—a story can tune them out entirely as it focuses in on the task at hand: *What does your protagonist have to confront in order to solve the problem you've so cleverly set up for her?* And that problem is what the reader is going to be hunting for from the get-go, because it's going to define everything that happens from the first sentence on.

What Rapidly Unraveling Situation Have You Plunked Me Into, Anyway?

Let's face it, we're all busy. Plus, most of us are plagued by that little voice in the back of our head constantly reminding us of what we really should be doing right now instead of whatever it is we're actually doing—especially when we take time out to do something as seemingly nonproductive as, um, read a novel. Which means that in order to distract us from the relentless demands of our immediate surroundings, a story has to grab our attention fast.[8] And, as neuroscience writer Jonah Lehrer says, nothing focuses the mind like surprise.[9] That means when we pick up a book, we're jonesing for the feeling that something out of the ordinary is happening. We crave the notion that we've come in at a crucial juncture in someone's life, and not a moment too soon. What intoxicates us is the hint that not only is trouble brewing, but it's longstanding and about to reach critical mass. This means that from the first sentence we need to catch sight of the breadcrumb trail that will lure us deeper into the thicket. I've heard it said that fiction (all stories, for that matter) can be summed up by a single sentence—All is not as it seems—which means that what we're hoping for in that opening sentence is the sense that something is about to change (and not necessarily for the better).

Simply put, we are looking for a reason to care. So for a story to grab us, not only must something be happening, but also there must be a consequence we can anticipate. As neuroscience reveals, what draws us into a story and keeps us there is the firing of our dopamine neurons, signaling that intriguing information is on its way.[10] This means that whether it's an actual event unfolding or we meet the protagonist in the midst of an internal quandary or there's merely a hint that something's slightly "off" on the first page, there has to be a ball already in play. Not the preamble to the ball. Not all the stuff you have to know to really understand the ball. The ball itself. This is not to say the first ball must be the main ball—it can be the initial ball or even a starter ball. But on that first page, it has to feel like the only ball and it has to have our complete attention.

For instance, how about this—the first paragraph of Caroline Leavitt's *Girls In Trouble*—for a ball in play?

> Sara's pains are coming ten minutes apart now. Every time one comes, she jolts herself against the side of the car, trying to disappear. Everything outside is whizzing past her from the car window because Jack, her father, is speeding, something she's never seen him do before. Sara grips the armrest, her knuckles white. She presses her back against the seat and digs her feet into the floor, as if any moment she will fly from the car. *Stop*, she wants to say. *Slow down. Stop*. But she can't form the words, can't make her mouth work properly. Can't do anything except wait in terror for the next pain. Jack hunches over the wheel, beeping his horn though there isn't much traffic. His face is reflected in the rearview mirror, but he doesn't look at her. Instead, he can't seem to keep himself from looking at Abby, Sara's mother, who is sitting in the back with Sara. His face is unreadable.[11]

Trouble brewing? Yep. Longstanding trouble? At least nine months, probably longer. Can't you feel the momentum? It pulls you forward, even as it grounds you in the unfolding moment. You want to know not only what happens next but also what led to what's happening right now. Who's the father? Was it consensual? Was she raped? Thus your curiosity is engaged, and you read on without consciously having made the decision to do so.

What Does *That* Mean?

As readers we eagerly probe each piece of information for significance, constantly wondering, "What is this meant to tell me?" It's said people can go forty days without food, three days without water, and about thirty-five seconds without finding meaning in something—truth is, thirty-five seconds is an eternity compared to the warp speed with which our subconscious brain rips through data. It's a biological imperative: we are always on the hunt for meaning—not in the metaphysical "What is the true nature of reality?" sense but in the far more primal, very specific sense of: *Joe left without his usual morning coffee; I wonder why? Betty is always on time; how come she's half an hour late? That annoying dog next door barks its head off every morning; why is it so quiet today?*

We are always looking for the *why* beneath what's happening on the surface. Not only because our survival might depend on it, but because it's exhilarating. It makes us feel something—namely, curiosity. Having our curiosity piqued is visceral. And it leads to something even more potent: the anticipation of knowledge we're now hungry for, a sensation caused by that pleasurable rush of dopamine. Because being curious is necessary for survival (*What's that rustling in the bushes?*), nature encourages it. And what better way to encourage curiosity than to make it feel good? This is why, once your curiosity is roused as a reader, you have an emotional, vested interest in finding out what happens next.

And bingo! You feel that delicious sense of urgency (hello dopamine!) that all good stories instantly ignite.

Do You Want an Interpreter with That?

So what happens when you can't anticipate what might happen next, when you can't even make sense of what's happening *now*? Usually you decide to find something else to read, pronto. I've often thrown up my hands in frustration when reading a well-intentioned manuscript, wishing it came with an interpreter. I could feel the author's burning intent; I knew she was trying to tell me something important. Trouble was, I had no idea what.

Think of how exasperating it is in the real world when someone begins a long rambling story:

> Did I tell you about Fred? He was supposed to come over last night, but it was raining, and like a dolt I forgot to shut my windows and my new couch got soaked. I paid a fortune for it. I'm worried that now it'll mildew like the old clothes in my grandma's attic. She's so dingy, but I can't blame her. She's over a hundred. I hope I have her genes. She was never sick a day in her life, but lately I've begun to wonder because my joints hurt every time it rains. Boy, they sure were aching last night while I was waiting for Fred. . . .

By now you're probably nervously jiggling your foot and thinking, *What are you talking about and why should I care?* That is, if you're still listening. It's the same with the first page of a story. If we don't have a sense of what's happening and why it matters to the protagonist, we're not going to read it. After all, have you ever gone into a bookstore, pulled a novel off the shelf, read the first few pages and thought, *You know, this is kind of dull, and I don't really care about these people, but I'm sure the author tried really hard and probably has something important to say, so I'm going to buy it, read it, and recommend it to all my friends?*

Nope. You're beautifully, brutally heartless. I'm betting you never give the author's hard work or good intentions a second thought. And that's as it should be. As a reader, you owe the writer absolutely nothing. You read their book solely at your own pleasure, where it stands or falls on its own merit. If you don't like it, you simply slip it back onto the shelf and slide out another.

What are you hunting for on that first page? Are you consciously analyzing each sentence one by one? Are you aware of what triggers the finely calibrated tipping point when you decide to either read the book or look for another? Of course not. That is, not consciously. In the same way you don't have to think about which muscles you need to move in order to blink, choosing a book is a perfectly coordinated reflex orchestrated by your cognitive unconscious. It's muscle memory—except in this case, the "muscle" in question is the brain.

Okay, let's say that the first sentence has indeed grabbed you. What's next?

What Is This Story About?

The unspoken question that's now bouncing around in your brain is this: *What is this book about?*

Sounds like a big question. It is, which is why we'll be exploring it in depth in the next chapter. So *can* you answer it on the first page? Rarely. After all, when you meet someone new, can you know everything there is to know about that person on the first date? Absolutely not. Can you feel like you do? Absolutely. Story, likewise. And to that end, here are the three basic things readers relentlessly hunt for as they read that first page:

1. Whose story is it?

2. What's happening here?

3. What's at stake?

Let's examine these three elements and how they work in tandem to answer the question.

WHOSE STORY IS IT?

Everyone knows a story needs a main character, otherwise known as the protagonist—even ensemble pieces tend to have one central character. No need to discuss it, right? But here's something writers often don't know: in a story, what the reader feels is driven by what the protagonist feels. Story is visceral. We climb inside the protagonist's skin and become sensate, feeling what he feels. Otherwise we have no port of entry, no point of view through which to see, evaluate, and experience the world the author has plunked us into.

In short, without a protagonist, everything is neutral, and as we'll see in chapter 3, in a story (as in life) there's no such thing as neutral. Which means we need to meet the protagonist as soon as possible—hopefully, in the first paragraph.

WHAT'S HAPPENING HERE?

It stands to reason, then, that something must be happening—beginning on the first page—that the protagonist is affected by. Something that gives us a glimpse of the "big picture." As John Irving once said, "Whenever possible, tell the whole story of the novel in the first sentence."[12] Glib? Yeah, okay. But a worthy goal to shoot for.

The big picture cues us to the problem the protagonist will spend the story struggling with. For instance, in a classic romantic comedy it's *Will boy get girl?* Thus we gauge every event against that one question.

Does it help him get closer to her or does it hurt his prospects? And, often, is she really the right girl for him?

Which brings us to the third thing that readers are hunting for on that first page, the thing that, together with the first two, ignites the all-important sense of urgency:

WHAT'S AT STAKE?

What hangs in the balance? Where's the conflict? Conflict is story's lifeblood—another seeming no-brainer. But there's a bit of helpful fine print that often goes unread. We're not talking about just any conflict, but conflict *that is specific to the protagonist's quest*. From the first sentence, readers morph into bloodhounds, relentlessly trying to sniff out what is at stake here and how will it impact the protagonist. Sure, they're not quite certain what his or her quest is yet, but that's what they're hoping to find out by asking these questions. Point being—something must be at stake, beginning on the first page.

The Obvious Question

Can all three of these things be there on the first page? You bet. In 2007, literary theorist Stanley Fish published an editorial in the *New York Times* that answers just that question. He was rushing through an airport with only minutes to spare and nothing to read. He decided to dash into the bookstore and choose a book based solely on its first sentence. Here is the winner, from Elizabeth George's *What Came Before He Shot Her*:

> "Joel Campbell, eleven years old at the time, began his descent into murder with a bus ride."

Imagine that: all three questions were answered in a single sentence.

1. **Whose story is it?** Joel Campbell's.

2. **What's happening here?** He's on a bus, which has somehow triggered what will result in murder. (Talk about "all is not as it seems"!)

3. **What is at stake?** Joel's life, someone else's life, and who knows what else.

Who wouldn't read on to find out? The fact that Joel is going to be involved in a murder not only gives us an idea of what the book is about, it provides the context—the yardstick—by which we are then able to measure the significance and emotional meaning of everything that "comes before he shoots her."

Which is important, because after that first sentence, the novel follows the hapless, brave, poverty-stricken Joel through inner-city London for well over six hundred pages before the murder in question. But along the way we're riveted, weighing everything against what we know is going to happen, always wondering if *this* is the event that will catapult Joel into his fate, and analyzing why each twist and turn pushes him toward the inevitable murder.

Here's something even more interesting: without that opening sentence, *What Came Before He Shot* her would be a very, very different story. Things would happen, but we'd have no real idea what they were building toward. So, regardless of how well written it is (and it *is*), it wouldn't be nearly as engaging. Why?

Because, as neuropsychiatrist Richard Restak writes, "Within the brain, things are always evaluated within a specific context."[13] It is context that bestows meaning, and it is meaning that your brain is wired to sniff out. After all, if stories are simulations that our brains plumb for useful information in case we ever find ourselves in a similar situation, we sort of need to know what the situation *is*.

By giving us a glimpse of the big picture, George provides a yardstick that allows us to decode the meaning of everything that befalls Joel. Such yardsticks are like a mathematical proof—they let the reader anticipate what things are adding up to. Which makes them even more useful for the intrepid writer, because a story's yardstick mercilessly reveals those passages that don't seem to add up at all, unmasking them as the one thing you want to banish from your story at all costs.

The Boring Parts

Elmore Leonard famously said that a story is real life with the boring parts left out. Think of the boring parts as anything that doesn't relate to or affect your protagonist's quest. Every single thing in a story—including subplots, weather, setting, even tone—must have a clear impact on what the reader is dying to know: *Will the protagonist achieve her goal? What will it cost her in the process? How will it change her in the end?* What hooks us, and keeps us reading, is the dopamine-fueled desire to know what happens next. Without that, nothing else matters.

But what about stunning prose? you may ask. *What about poetic imagery?*

Throughout this book we'll be doing a lot of myth-busting, exploring why so many of the most hallowed writing maxims are often more likely to lead you in the wrong direction than the right. And this, my friends, is a great myth to start with.

MYTH: Beautiful Writing Trumps All

REALITY: Storytelling Trumps Beautiful Writing, Every Time

Few notions are more damaging to writers than the popular belief that writing a successful story is a matter of learning to "write well." Who could argue with that? It sounds so logical, so obvious. What would the alternative be—learning to write poorly? Ironically, writing poorly can be far less damaging than you'd think. That is, *if you can tell a story*.

The problem with this, along with numerous other writing myths, is that it misses the point. In this case, "writing well" is taken to mean the use of beautiful language, vibrant imagery, authentic-sounding dialogue, insightful metaphors, interesting characters, and a whole lot of really vivid sensory details dribbled in along the way.

Sounds pretty good, doesn't it? Who'd want to read a novel without it?

How about the millions of readers of *The Da Vinci Code*? Regardless of how beloved his books may be, no one says author Dan Brown is a great writer. Perhaps most succinct, and scathing, is fellow author Philip Pullman's assessment that Brown's prose is "flat, stunted and ugly," and that his books are full of "completely flat and two-dimensional characters . . . talking in utterly implausible ways to one another."[14]

So why is *The Da Vinci Code* one of the best-selling novels of all time? Because, from the very first page, readers are dying to know what happens next. And that's what matters most. A story must have the ability to engender a sense of urgency from the first sentence. Everything else—fabulous characters, great dialogue, vivid imagery, luscious language—is gravy.

This is not to disparage great writing in any way. I love a beautifully crafted sentence as much as the next person. But make no mistake: learning to "write well" is not synonymous with learning to write a story. And of the two, writing well is secondary. Because if the reader doesn't want to know what happens next, so what if it's well written? In the trade, such exquisitely rendered, story-less novels are often referred to as a beautifully written *"Who cares?"*

Now that we know what hooks a reader on the first page, the question is, how do you craft a story that actually does it? Like everything in life, it's easier said than done, which is why it's the question we'll spend the rest of the book answering.

CHAPTER 1: CHECKPOINT ✔️

Do we know whose story it is? There must be someone through whose eyes we are viewing the world we've been plunked into—aka the protagonist. Think of your protagonist as the reader's surrogate in the world that you, the writer, are creating.

Is something happening, beginning on the first page? Don't just set the stage for later conflict. Jump right in with something that will affect the protagonist and so make the reader hungry to find out what the consequence will be. After all, unless something is already happening, how can we want to know what happens next?

Is there conflict in what's happening? Will the conflict have a direct impact on the protagonist's quest, even though your reader might not yet know what that quest is?

Is something at stake on the first page? And, as important, is your reader aware of what it is?

Is there a sense that "all is not as it seems"? This is especially important if the protagonist isn't introduced in the first few pages, in which case it pays to ask: Is there a growing sense of focused foreboding that'll keep the reader hooked until the protagonist appears in the not-too-distant future?

Can we glimpse enough of the "big picture" to have that all-important yardstick? It's the "big picture" that gives readers perspective and conveys the point of each scene, enabling them to add things up. If we don't know where the story is going, how can we tell if it's moving at all?

2

HOW TO ZERO IN ON YOUR POINT

COGNITIVE SECRET:

*When the brain focuses its full attention on something,
it filters out all unnecessary information.*

STORY SECRET:

*To hold the brain's attention, everything in a story
must be there on a need-to-know basis.*

Stick to the point.

—W. SOMERSET MAUGHAM

HERE'S A DISCONCERTING THOUGHT: marketers, politicians, and tel-evangelists know more about story than most writers. This is because, by definition, they start with something writers often never even think about—the point their story will make. Armed with that knowledge, they then craft a tale in which every word, every image, every nuance leads directly to it.

Look around your house. Chances are you bought just about everything you see (even Fido) because while you weren't looking, a clever story snuck in and persuaded you to. It's not that you're easy to boss around, but a well-crafted story speaks first to your cogni-tive unconscious[1]—which marketers hope will then translate it into something conscious, like, *It may be midnight, but I really do deserve a Big Mac. Gee, she looks so happy; I wonder if I can get my doctor to prescribe that pill. It'd sure be fun to have a beer with that guy, I think I'll vote for him.*

Scary, huh?

So to take back some of that power, writers would do well to embrace this counterintuitive fact: the defining element of a story is something that has little to do with writing. Rather, it underlies the story itself and is what renowned linguist William Labov has dubbed "evaluation" because it allows readers to evaluate the meaning of the story's events. Think of it as the "So what?" factor.[2] It's what lets read-ers in on the point of the story, cluing them in to the relevance of everything that happens in it. Put plainly, it tells them what the story is about. As literary scholar Brian Boyd so aptly points out, a story with no point of reference leaves the reader with no way of determining what information matters: is it "the color of people's eyes or their socks? The shape of their noses or their shoes? The number of syllables in their name?"[3]

Thus your first job is to zero in on the point your story is making. The good news is that this is one of the few things that can actually cut down on time spent rewriting. Why? Because from the get-go it allows you to do for your story what your cognitive unconscious automatically does for you: filter out unnecessary and distracting information.[4]

To that end, in this chapter we'll explore how weaving together the protagonist's issue, the theme, and the plot keeps a story focused; what theme really means and how it defines your story; and the ways in which plot can get in your way. Then we'll put these principles through a test run, focusing on that literary classic, *Gone with the Wind*.

A Story Versus Stuff That Happens

A story is designed, from beginning to end, to answer a single over-arching question. As readers we instinctively know this, so we expect every word, every line, every character, every image, every action to move us closer to the answer. Will Romeo and Juliet run off together? Will Scarlett realize Rhett's the man for her before it's too late? Will we find out enough about Charles Foster Kane to know what the hell Rosebud means?

Thus it would seem that when you're writing a story, defining what it's about should be simple—obvious, almost—yet it often proves to be maddeningly elusive. Despite our best intention, the narrative meanders, spending way too much time wandering aimlessly down back roads. So in the end, although a lot of interesting events take place, they don't add up to anything. No question is asked, let alone answered. The story is so full of things the reader doesn't need to know that it has no focus, so it isn't really a story. It's just a collection of things that happen.

Stories that lack focus often aren't about anything at all. Sounds impossible, doesn't it? But I can't tell you how many manuscripts I've read where if someone asked, "What's it about?" my only answer would be, "It's about three hundred pages." As one editor put it, "If you can't summarize your book in a few sentences, rewrite *the book* until you can."

I agree. Years of reading query letters, synopses, and countless manuscripts and screenplays have taught me that writers who can't sum up the story they're telling in a clearly focused, intriguing sentence or two probably haven't written a clearly focused, intriguing story. It wasn't a lesson that came easy. I'd read a summary that seemed promising but was jumbled and a little disjointed and think, *Hey, the ability to write a good story is very different from the ability to write a good summary.* So I'd start reading the manuscript. I rarely got far, however, because it usually turned out that the summary did present an accurate picture of the story, which was itself disjointed and jumbled.

Here are just a few telltale signs that a story is going off the rails:

- We have no idea who the protagonist is, so we have no way to gauge the relevance or meaning of anything that happens.

- We know who the protagonist is, but she doesn't seem to have a goal, so we don't know what the point is or where the story is going.

- We know what the protagonist's goal is, but have no clue what inner issue it forces him to deal with, so everything feels superficial and rather dull.

- We know who the protagonist is and what both her goal and her issue are, but suddenly she gets what she wants, arbitrarily changes her mind, or gets hit by a bus, and now someone else seems to be the main character.

- We're aware of the protagonist's goal, but what happens doesn't seem to affect him *or* whether he achieves it.

- The things that happen don't affect the protagonist in a believable way (if at all), so not only doesn't she seem like a real person, but we have no idea why she does what she does, which makes it impossible to anticipate what she'll do next.

All these problems have the same effect on the reader's brain: not only does the dopamine surge we felt when we started reading dry up, but the part of our brain always busily comparing the reward we expected with what we actually got lets us know it is not pleased. In short, we feel frustrated.[5] This is evidence that the author hasn't zeroed in on the essence of the story she's telling, so even though it may be brimming with exquisite prose, it feels directionless and uninvolving. It doesn't take a neuroscientist to tell us what happens next. We stop reading. End of story.

The Crucial Importance of Focus

What was missing in all those failed manuscripts is focus. Without it, the reader has no way to gauge the meaning of anything, and since we're wired to hunt for meaning in everything—well, you do the math. A story without focus has no yardstick.

So, what is this thing called focus? It's the synthesis of three elements that work in unison to create a story: the protagonist's issue, the theme, and the plot. The seminal element—the protagonist's issue—stems from something we mentioned in the last chapter: the story question, which translates to the protagonist's goal. But remember what we said? The story isn't about whether or not the protagonist achieves her goal per se; it's about what she has to overcome *internally* to do it. This is what drives the story forward. I call it the protagonist's issue.

The second element, the theme, is what your story says about human nature. Theme tends to be reflected in how your characters treat each other, so it defines what is possible and what isn't in the world the story unfolds in. As we'll see, it's often what determines whether the protagonist's efforts will succeed or fail, regardless of how heroic she is.

The third element is the plot itself—the events that relentlessly force the protagonist to deal with her issue as she pursues her goal, no matter how many times she tries to make an end run around her issue along the way.

Taken together, these three elements give a story focus, telling readers what it's about and allowing them to interpret the events as they unfold and thus anticipate where it's heading. This is crucial because "minds exist to predict what will happen next."[6] It's their *raison d'être*—the better to keep us on this earthly plane as long as humanly possible. We love to figure things out and we don't like being confused. For writers, focus is of utmost importance as well: the first two elements (the protagonist's issue and the theme) are the lens through which we determine what the events (the plot) will be.

How do they do this? By setting the story's parameters and zeroing in on the particular aspect of the protagonist's life it will chronicle. After all, our characters live their lives 24/7 just like we do; they eat, sleep, argue with insurance companies, get annoyed when the Internet goes down, veg out in front of the TV, and spend time trying to remember whether that dentist appointment is Tuesday or Thursday. Would you put all of that in a story? Of course not. Instead, you cherry-pick events that are relevant to the story question and construct a gauntlet of challenge (read: the plot) that will force the protagonist to put his money where his mouth is. Think baptism by ever-escalating fire.

Done right, we have another mathematical proof, a concrete frame of reference against which everything that happens is measured. After all, this is exactly how our brain processes information when we're confronted with a sticky situation in real life. As neuroscientist Antonio Damasio demonstrates, this is what literature is modeled on:

> Suppose you are sitting down for a cup of coffee at a restaurant
> to meet with your brother, who wishes to discuss your parents'

inheritance and what is to be done with your half sister, who has been acting strangely. You are very present and in the moment, as they say in Hollywood, but now you are also transported, by turns, to many other places, with many other people besides your brother, and to situations that you have not experienced yet that are the products of your informed and rich imagination. . . . You are busily all over the place and at many epochs of your life, past, and future. But you—the *me* in you—never drops out of sight. *All of these contents are inextricably tied to a singular reference. Even as you concentrate on some remote event, the connection remains. The center holds.* This is big-scope consciousness, one of the grand achievements of the human brain and one of the defining traits of humanity. . . . This is the kind of consciousness illustrated by novels, films, and music. . . .[7] (Italics mine.)

In other words, the center—here, how the question of what to do about said inheritance affects our friend in the restaurant—is the singular reference that everything else relates to. If this *were* a story, our friend would have an internal issue he would need to work through in order to navigate this inherently thorny situation. Would he be successful? That's where the theme comes in.

But What *Is* Theme, Exactly?

There's a lot of talk about what theme is, and how it's revealed, which can result in esoteric discussions capable of parsing it down to the thematic use of margarine as a metaphor for innocence lost. Happily, theme actually boils down to something incredibly simple:

- What does the story tell us about what it means to be human?

- What does it say about how humans react to circumstances beyond their control?

Theme often reveals your take on how an element of human nature—loyalty, suspicion, grit, love—defines human behavior. But the real secret to theme is that it's not general; that is, the theme wouldn't be "love" per se—rather, it would be a very specific point you're making about love. For instance, a love story can be sweet and lyrical, revealing that people are good eggs after all; it can be hard-nosed and edgy, revealing that people are intense and quirky; it can be cynical and manipulative, revealing that people are best avoided, if possible.

Knowing the theme of your story in advance helps, because it gives you a gauge by which to measure your characters' responses to the situations they find themselves in. They'll be kind, gruff, or conniving depending on the universe you have created for them. This, then, affects how the story question is resolved, because it governs the type of resistance the protagonist will meet along the way. In a loving universe, she may discover that, with a little gumption, she'll find her true love. In an impersonal universe, she'll find no one she can really relate to, and in a cruel universe, she'll end up married to Hannibal Lecter.

What's Your Point?

Theme often reveals the point your story is making—and all stories make a point, beginning on page one. But that doesn't mean you have to hit readers over the head with it.

Think about advertising. An ad's goal is to deliver a very specific punch without letting us know exactly how it's doing it, even though when it comes to ads, we *know* what their intention is: to get us to buy the product. As corporate consultants Richard Maxwell and Robert Dickman say in their book *The Elements of Persuasion,* "For those of us whose business depends on being able to persuade others—which is all of us in business—the key to survival is being able to cut through all the clutter and make the sale. The good news is that the secret of selling is what it has always been—a good story."[8]

Knowing your story's point is what helps you cut through all the clutter.

Not that you're as calculating as an advertising executive or that your story has so literal a purpose, which is why writers often have to stop and think about what it is they're trying to say and what point their story is making. It's crucial, because the instant a reader opens your book, his cognitive unconscious is hunting for a way to make life a little easier, see things a little clearer, understand people a bit better.[9] So why not take a second to ask yourself, *What is it I want my readers to walk away thinking about? What point does my story make? How do I want to change the way my reader sees the world?*

Don't Bury Your Story in an Empty Plot

It's not surprising that of the three elements that combine to create focus, writers often dote on only one of them—the plot. Because it's the element that, by definition, is the vehicle for the other two, it's easy to forget they're there. Trouble is, without them the plot ends up an empty vessel—things happen, but no one is really affected by them, especially the reader. This brings us to another common myth in need of shattering:

MYTH: The Plot Is What the Story Is About

REALITY: A Story Is About How the Plot *Affects* the Protagonist

While thus far it's been implied, it helps to say it flat out: plot is not synonymous with story. Plot *facilitates* story by forcing the protagonist to confront and deal with the issue that keeps him from achieving his goal. The way the world treats him, and how he reacts, reveals the theme. So at the end of the day, what the protagonist is forced to learn as he navigates the plot is what the story is about. It's important to always keep this in mind since the plot, when taken by itself, can suggest that a story is about one thing when in reality, it's about something else.

A great example of this can be found in the movie *Fracture*—which, like many movies, makes a great case study for overarching story concepts. Why? Because story-wise, film is often a simpler, more straightforward medium than prose (not to mention that people are far more likely to have seen the same movies than to have read the same books). In *Fracture*, we don't meet the protagonist, Willy Beachum, for a full seventeen minutes. Until then we assume the protagonist is Ted Crawford, whom we watch mortally wound his wife in cold blood a few minutes into the film. We believe the story will be about whether or not Crawford goes to jail for it, and in fact, that *is* what the plot chronicles.

But it's *not* what the story is about. Instead, *Fracture* is about whether Beachum—a hotshot prosecutor who gets the case just as he's about to leave the public sector and take a cushy job in a white-shoe law firm—will end up compromising his integrity by selling out, or whether he'll fight the good fight and stay on in the prosecutor's office (*adios*, dreams of wealth and prestige). Thus the plot—Crawford and his trial—occurs solely to test Beachum's moral fiber. So although Beachum doesn't appear until almost twenty minutes into the film, everything that happens up to that moment occurs, story-wise, solely to put him to the test.

In other words, even when the protagonist doesn't appear on the first page, everything that happens before he shows up must occur with a clear eye toward how it will affect him when he finally ambles in. This is not to say readers will be aware of it until then. How could they be? After all, in *Fracture* we have no idea the story isn't about Crawford until Beachum makes his entrance. But the writers knew. So they made sure everything Crawford did would come back to test Beachum's resolve (and "test" it not in the general sense, but in a very specific, focused way). Because each of Crawford's very calculated actions was devised to challenge Beachum's view of himself, of the world, and of his place in it. As the story progresses, these actions "fracture" his otherwise cocksure, self-absorbed persona, allowing something far more meaningful, and gritty, to emerge.

What does *Fracture* have to say about the human condition? That at the end of the day, integrity is worth far more than wealth, even if it means that you have to live out of your car for a while. Ah, but how is this message delivered? In the guise of a compelling, fast-moving plot that allows us to burrow deeply into Beachum's skin as he wrestles with what is thrown at him. Thus we have a bird's-eye view of the battle between the protagonist and the plot, which we'll be discussing in more detail a little further on.

Theme: The Keys to the Universe

Since theme is the underlying point the narrative makes about the human experience, it's also where the universal lies. The universal is a feeling, emotion, or truth that resonates with us all. For instance, "the raw power of true love" is something everyone (okay, almost everyone) can tap into, whether the story is about a saloon owner in Casablanca, a mermaid under the sea, or a knight in Arthur's court. The universal is the portal that allows us to climb into the skin of characters completely different from us and miraculously feel what they feel.

Given the primacy of the universal, it's ironic that only when embodied in the *very specific* does a universal become accessible, as we'll explore in depth in chapter 6. In the abstract, universals are so vast they're impossible to wrap your mind around. It's only when expressed through the flesh-and-blood reality of a story, that we're able to experience a universal one-on-one, and so *feel* it.

The Pulitzer Prize–winning novel *Olive Kitteridge* offers a simple, sublime example. Its theme is how we bear loss, and author Elizabeth Strout has said that she hopes her readers "feel a sense of awe at the quality of human endurance."[10] In the following passage, a mundane moment triggers a memory that is utterly gripping because it taps into a universal that, I'd venture to say, everyone has experienced and yet rarely found the words to express:

> She was glad she had never left Henry. She'd never had a friend as loyal, as kind, as her husband.
>
> And yet, standing behind her son, waiting for the traffic light to change, she remembered how in the midst of it all there had been times when she'd felt a loneliness so deep that once, not so many years ago, having a cavity filled, the dentist's gentle turning of her chin with his soft fingers had felt to her like a tender kindness of almost excruciating depth, and she had swallowed with a groan of longing, tears springing to her eyes.[11]

In that very specific memory—the dentist's fleeting, workaday touch—an otherwise ineffable feeling of existential loneliness is made manifest, as palpable as if it had happened to us—because, as we'll see in chapter 4, as far as our brain is concerned, it actually has.

By filtering her story through the thematic lens of loss and human endurance, Strout was able to pluck an otherwise random moment from Olive's life and use it to give us insight into how Olive sees the world, and at the same time provide a visceral glimpse of the cost of being human.

Theme and Tone: It's Not What You Say but How You Say It

If theme is one of the most powerful elements of your story, it's also one of the most invisible. You didn't "see" the theme anywhere in Strout's passage, did you? It wasn't spelled out, wasn't referenced, but it was there, all the same. It's like tone of voice, which often says more than the words themselves. In fact, sometimes tone says the exact opposite of what the words are saying, as anyone who's ever been in a long-term relationship can attest.

Your story's tone reflects how you see your characters and helps define the world you've set them loose in. Tone is often how theme is conveyed, by cueing your readers to the emotional prism through which you want them to view your story—like a soundtrack in a movie. It's another way of sharpening your focus, highlighting what your reader really needs to know.

For instance, the tone in a romance novel lets us know that, although big things will definitely go wrong, nothing genuinely damaging will ever happen, so we can safely relax into the story, secure in the knowledge that love is not only capable of saving the day, but actually will. Whereas in a novel like *What Came Before He Shot Her,* from the first sentence, the tone implies the exact opposite, though it doesn't come right out and tell us so. Instead, tone makes us feel it, by evoking a particular mood. Tone belongs to the author; mood to the reader.

In other words, your theme begets the story's tone, which begets the mood the reader feels. Mood is what underlies the reader's sense of what is possible and what isn't in the world of your story, which brings us back to the point your story is making as reflected in its theme— *reflected* being the key word. Because as crucial as theme is, it's never stated outright; it's always implied. Movies and books that put theme first and story second tend to break the cardinal (although often grievously misunderstood, as we'll see in chapter 7) rule of writing, "Show, don't tell." It's the story's job to show us the theme, not the theme's job to tell us the story—especially since theme is a rotten storyteller and, when left to its own devices, is much more interested in telling us what to think than in simply presenting the evidence and letting us make up our own mind. Unchecked, theme is a bully, a know-it-all. And no one likes to be told what to do, which is why reverse psychology works so well. What this means is that the more passionate you are about making your point, the more you have to trust your story to convey it. As Evelyn Waugh says, "All literature implies moral standards and criticisms, the less explicit the better."[12]

Besides, did you ever go into a bookstore saying to yourself, *What I'd really like is a book about survival and how catastrophes bring out the gumption in some and not in others?*[13] Or *I'm dying to curl up with a good book that traces the defects of society back to the defects of human nature?*[14] Or *What I'm so in the mood for is a book that is a metaphor for Latin America?*[15] I don't think so. Which isn't to say that you might not leave with *Gone with the Wind, Lord of the Flies,* or *One Hundred Years of Solitude,* whose authors, when pressed, described their themes as such.

But wait: aren't there more themes in each of those books? Probably. In fact, a simple Internet search will turn up myriad suggested themes for each title—some of which would no doubt stun, if not infuriate, their authors. But they are mostly secondary themes. What we're talking about is the main theme—the one you, the writer, choose, rather than the ones scholars will later foist upon you so graduate students can endlessly debate them in small, earnest seminars.

Gone with the Wind: A Case Study

To better understand how to use focus to define what your book is about—thus creating a yardstick by which to filter out all unnecessary information—let's look at the most accessible of the three books just mentioned: *Gone with the Wind.* In the past some have dismissed *Gone with the Wind* as a trite, melodramatic potboiler, nothing more than "popular fiction." But no one can deny its power as a spellbinding page-turner. And here's the shocker: in 1937 it won the Pulitzer Prize. It also happened to be the bestselling novel of all time until it was surpassed in 1966 by *Valley of the Dolls*—which somehow the Pulitzer committee overlooked.

First, let's take a good look at the theme of *Gone with the Wind* according to author Margaret Mitchell in an interview with her publisher in 1936:

> If it has a theme it is that of survival. What makes some people able to come through catastrophes and others, apparently just as able, strong and brave, go under? It happens in every upheaval. Some people survive; others don't. What qualities are in those who fight their way through triumphantly that are lacking in those who go under? I only know that the survivors used to call that quality "gumption." So I wrote about the people who had gumption and the people who didn't.[16]

As Scarlett fights, schemes, manipulates, struggles, and ultimately survives against all odds, the key ingredient is *gumption*. Fair enough. But is that the novel's main thematic focus? Does it drive Scarlett's reaction as calamity after calamity befall her? Is it the lens through which we watch the tale unfold? The secret ingredient that holds us fast, whether we can define it or not? It is.

What keeps us reading is the knowledge that Scarlett's headstrong will, her guts, her nerve—her gumption—is stronger than her need to conform to society's dictates. But we quickly learn that, as potent as her untempered gumption is, it's also capable of completely blinding her to what is in her best interest—which, as we'll soon see, is where her internal issue lies. We know what would make her the happiest. And we realize pretty quickly that chances are it's the last thing she'll do. Which raises the question: *What will she do instead? Will she ever wake up and realize what she truly wants?* And that's what keeps us reading.

But what about the other themes that run through the novel—for instance, the nature of love, the constraints of class structure, and of course, nineteenth century society's tightly corseted gender roles? Couldn't any one of them be the central theme? Good question. Here's the litmus test: *the central theme must provide a point of view precise enough to give us specific insight into the protagonist and her internal issue, yet be broad enough to take into account everything that happens* (again: the plot). Let's see what happens when I try to sum up *Gone with the Wind* with these other contenders. First, the nature of love:

> Set against a backdrop of the Civil War, *Gone with the Wind* is about a Southern belle whose misguided love for the wrong man blinds her to the one person who could give her what she wants.

It's not a bad description—if the book were solely a romance, with everything else merely "setting." But given the novel's scope, it's much too limiting.

Well, then, what about the way Scarlett disregards social norms?

> *Gone with the Wind* is about a Southern belle who bucks the societal tide in order to survive during the Civil War.

This one isn't bad either. That is, if you go in for the general. What societal tide, exactly? Buck it, how? Without any specifics, it's hard to get a real picture of . . . much of anything. Okay, what about class structure?

> *Gone with the Wind* is about how traditional class structure in the South gave way during the Civil War.

Sounds like nonfiction, doesn't it? And since nonfiction sells, and there are millions of Civil War buffs, this could be a bestseller—that is, until they realize it's really a steamy romance about a gutsy woman who ruthlessly bucks the societal tide. Of course, by then even the staunchest history buff might keep mum, too busy hoping against hope that Scarlett wakes the hell up and realizes that Rhett is the man for her before it's too damn late.

So, although this isn't to say that my descriptions wouldn't entice some readers, there is nothing in them that suggests a sprawling, steamy epic, and *Gone with the Wind* is nothing if not that. But when I begin with gumption—the notion Mitchell used as her defining theme—it's another story:

> *Gone with the Wind* is about a headstrong Southern belle whose unflinching gumption causes her to spurn the only man who is her equal, as she ruthlessly bucks crumbling social norms in order to survive during the Civil War.

Aha! While my description of *Gone with the Wind* might not be there yet, we've hit on something well worth mentioning. One way to help identify a story's defining theme is to ask yourself: is it possible to filter the story's other themes through it? In *Gone with the Wind,* Scarlett's gumption came first, so—for better or worse—it affects everything else: her love life, her refusal to be constrained by the mores of the day, and her insatiable need to take action when she doesn't get what she wants. Take action? Ah yes, the plot.

THE PROTAGONIST'S ISSUE VERSUS THE PLOT

As we know, it's the plot that puts the protagonist through his paces, presenting increasingly difficult obstacles that must be overcome if he's to get within grabbing distance of the brass ring.

But the plot's goal isn't simply to find out whether he snags that brass ring or not; rather, it's to force him to confront the internal issue that's keeping him from it in the first place. This issue is sometimes called the protagonist's "fatal flaw," and whether a deep-rooted fear, a stubborn misperception, or a dubious character trait, it's what he's been battling throughout and what he must finally overcome to have a clear shot at the last remaining obstacle. Ironically, once he overcomes it, he often realizes true success is vastly different from what, up to that very second, he thought it was. This is frequently the case in romantic comedies and is usually the moment when the big lug finally realizes that the beautiful, stuck-up, rich, thin girl he's been hell-bent on winning since the opening credits isn't *nearly* as loveable as the cute, cuddly, beautiful, thin middle-class girl next door.

Not so with Scarlett.

Scarlett's fatal flaw is self-absorption, which when harnessed to her unstoppable gumption, makes her vulnerable in a way she cannot see. But we can. And so we're rooting for her not only to survive, but also to gain enough self-awareness to keep herself from throwing out the baby with the bathwater. Does she? Almost, but she's a day late and a dollar short. Which is why when the book ends, unlike Rhett, we do give a damn.

SCARLETT'S SPECIFIC GOAL—WHAT DOES SCARLETT *REALLY* WANT?

But wait; it still feels like something's missing in our description of the novel. Sure, fatal flaw or not, Scarlett wants to survive. But don't we all? Indeed we do, which makes survival, in and of itself, generic—one of those abstract universals. In other words, the same would be true of everyone, so it doesn't tell us a thing about Scarlett herself and adds nothing to the story. The question is: *What does survival mean to Scarlett?* Plot-wise (that is, on the corporeal plane where the action unfolds) this translates to: *What does Scarlett need in order to feel she's survived what life has thrown at her?* The answer is her family's plantation, Tara. Meaning, *land*. As her father tells her early on, "Land is the only thing in the world that amounts to anything. . . ." Land is what ties you to your past and makes you who you are. Without it, you are nothing. This becomes Scarlett's benchmark, the thing that she's sure will prove she's survived.

Is she right? *Is* land what ties you to your past and makes you who you are? God, let's hope not. This is why Scarlett emerges both a successful and a tragic figure. And why her blindness to what she truly wants—caused by her fatal flaw—is understandable, rather than annoying or, worse, hair-pullingly frustrating to the reader. Readers are a surprisingly accepting lot when it comes to willfully blind protagonists, provided they understand the reason for their blindness. This is often exactly what such stories are about: why *would* a person work overtime to stay blind to something that is painfully clear to everyone else? In fact, sometimes the "aha!" moment belongs to the reader rather than the protagonist. It's the epiphany that comes of realization that not only isn't the protagonist going to change, but for the first time we grasp the full weight of what the self-imposed blindness is protecting her from.

So, getting back to Scarlett, let's add a clause to our description:

> *Gone with the Wind* is about a headstrong southern belle whose unflinching gumption causes her to spurn the only man who is her equal, as she ruthlessly bucks crumbling social norms in order to survive during the Civil War by keeping the one thing she mistakenly believes matters most: her family estate, Tara.

Focus, anyone? How's that for a mini-outline! We've taken the theme—survival driven by gumption—harnessed it to Scarlett's issue, and then run them both through the hurdles the plot lays out for her. By synthesizing the theme, Scarlett's internal issue, and the plot, we've boiled a 1,024-page book down to its essence. In one (albeit long) sentence, we've provided enough of the big picture to give a clear idea of what the book is about.

Harnessing Focus: How to Keep Your Story on Track

While clearly this is a very handy method for defining what, exactly, your story is about once it's written, it can be even more helpful before you begin writing—or at whatever stage your story is at right now. It's never too late or too early, and it always helps. Knowing what the focus of your story is allows you to do for your story what your cognitive unconscious does for you: filter out everything extraneous, everything that doesn't matter. You can use it to test each proposed twist, turn, and character reaction for story relevance.

This isn't to say that once you begin writing you might not change your mind about the theme, the story question, or that the story might not unfold in a completely different direction than you anticipated. But—and here's another reason why figuring these things out first makes all the difference—if it *does* change, you'll recognize it and be able to adjust the narrative accordingly. How? Because you've mapped out where the story was headed, you can now use the same map to rechart your story's course. Don't forget: when a story shifts focus halfway through, it not only means it's now heading in a different direction; it also means that everything leading up to that spot has to shift as well. Otherwise, it's like boarding a plane bound for New York City that lands in Cincinnati instead. Talk about disorienting (not to mention that you've packed all the wrong clothes). The good news is that because you already have a map—something we'll develop in more depth in chapter 5—you know just where those changes need to be.

This will please your readers immensely. Since their implicit belief is that everything in a story is there on a need-to-know basis, the last thing you want is for them to continually trip over all the unnecessary info cluttering up your otherwise splendid story.

CHAPTER 2: CHECKPOINT

Do you know what the point of your story is? What do you want people to walk away thinking about? How do you want to change how they see the world?

Do you know what your story says about human nature? Stories are our way of making sense of the world, so each and every one tells us something about what it means to be human, whether the author does it on purpose or not. What is *your* story saying?

Do the protagonist's inner issue, the theme, and the plot work together to answer the story question? How can you tell? Ask yourself: Is my theme reflected in the way the world treats my protagonist? Does each plot twist and turn force my protagonist to deal with his inner issue, the thing that's holding him back?

Do the plot and theme stick to the story question? Remember, the story question will always be in the back of your reader's mind, and it is the responsibility of each theme-laced event to keep it there.

Can you sum up what your story is about in a short paragraph? One way to begin is to ask yourself how your theme shapes your plot. Put yourself through the paces just as we did with *Gone with the Wind*. It may be painful, but it'll pay off big time in the end.

3

I'LL FEEL WHAT HE'S FEELING

COGNITIVE SECRET:

Emotion determines the meaning of everything—
if we're not feeling, we're not conscious.

STORY SECRET:

All story is emotion based—
if we're not feeling, we're not reading.

Indeed, feelings don't just matter—they are what mattering means.

—DANIEL GILBERT, *Stumbling on Happiness*

MOST OF US WERE BROUGHT UP to believe that reason and emotion are polar opposites—with reason as the stalwart white hat and emotion as the sulky black hat. And let's not even discuss which gender was said to wear which hat. Reason, it was thought, sees the world as it is, while that irrational scamp, emotion, tries to undermine it. Uh-huh.

Turns out, as neuroscience writer Jonah Lehrer says, "If it weren't for our emotions, reason wouldn't exist at all."[1] Take that, Plato! This is illustrated by a sad story that, even sadder, its real-life protagonist doesn't see as sad at all. Because he can't—literally. Elliot, a patient of Antonio Damasio, had lost a small section of his prefrontal cortices during surgery for a benign brain tumor. Before his illness, Elliot held a high-level corporate job and had a happy, thriving family. By the time he saw Damasio, Elliot was in the process of losing everything. He still tested in the 97th percentile in IQ, had a high-functioning memory, and had no trouble enumerating each and every possible solution to a problem. Trouble was, he couldn't make a decision—from what color pen to use to whether it was more important to do the work his boss expected or spend the day alphabetizing all the folders in his office.[2]

Why? Because, as Damasio discovered, the damage to his brain left him unable to experience emotion. As a result, he was utterly detached and approached life as if everything in it was neutral. But wait, shouldn't that be a good thing? Now that emotion couldn't butt in and cloud Elliot's judgment, he'd be free to make rational decisions, right? I think you know where this is going. Without emotion, each option carried the exact same weight—everything really was six of one, half a dozen of the other.

Turns out, as cognitive scientist Steven Pinker notes, "Emotions are mechanisms that set the brain's highest-level goals."[3] Along with, apparently, every other goal, down to what to have for breakfast. Without emotions, Elliot had no way to gauge what was important and what wasn't, what mattered and what didn't.

It is exactly the same when it comes to story. If the reader can't *feel* what matters and what doesn't, then nothing matters, including finishing the story. The question for writers, then, is where do these feelings come from? The answer's very simple: the protagonist.

In this chapter we'll explore how to deftly weave in the most important, yet often overlooked, element of story—letting the reader know how your protagonist is reacting internally to everything that happens, as it happens. We'll decode the secret of conveying thoughts when writing in first or third person; expose the sins of editorializing; take a good look at how body language never lies; and rethink that bossy old saw, "Write what you know."

The Protagonist: You Feel Me?

When we're fully engaged in a story, our own boundaries dissolve. We become the protagonist, feeling what she feels, wanting what she wants, fearing what she fears—as we'll see in the next chapter, we literally mirror her every thought. It's true of books and it's true of movies, too. I remember in college walking home after seeing an old Katharine Hepburn movie. It didn't occur to me how deeply I'd been affected until I caught my own reflection mirrored in a darkened store window. Until that moment, I'd been Katharine Hepburn. Or, more precisely, Linda Seton in *Holiday*. Then all of a sudden I was me again, which definitely meant that Cary Grant was not waiting for me on board a ship about to set sail into a glorious future.

But at least for a few splendid minutes walking down Shattuck Avenue, I saw the world through Linda Seton's eyes. It was visceral, and it felt like a gift—because my worldview had shifted. Linda was the black sheep of her family, and so was I. She'd fought tradition, regardless of the consequences, and even though she spent years in the proverbial attic, in the end, she triumphed. Maybe I could too. My step was lighter walking home than when I left for the theater.

This is a gift that so many of the manuscripts I've since read didn't quite bestow, because the author had fallen prey to a very common pitfall, one that in essence rendered their protagonist off-limits to the reader. They had mistaken the story for what happens in it. But as we've learned, the real story is how what happens affects the protagonist, and what she does as a result.

This means that *everything* in a story gets its emotional weight and meaning based on how it affects the protagonist. If it doesn't affect her—even if we're talking birth, death, or the fall of the Roman Empire—it is completely neutral. And guess what? Neutrality bores the reader. If it's neutral, it's not only beside the point, it detracts from it.

That's why in every scene you write, the protagonist must react in a way the reader can see and understand in the moment. This reaction must be specific, personal, and have an effect on whether the protagonist achieves her goal. What it can't be is dispassionate objective commentary.

Readers intuitively know what neuroscientists have discovered: everything we experience is automatically coated in emotion. Why? It's our version of a computer's ones and zeros, and it's based on a single question: *Will it hurt me, or will it help me?*[4] This humble equation underlies every aspect of our rich, elegant, complex, and ever-changing sense of self, and how we experience the world around us. According to Damasio, "No set of conscious images of any kind on any topic ever fails to be accompanied by an obedient choir of emotions and consequent feelings."[5] If we're not feeling, we're not breathing. A neutral protagonist is an automaton.

How to Catapult the Reader into Your Protagonist's Skin

When your protagonist's reaction is up close and personal, it catapults us into his skin, where we become "sensate," feeling what he feels, and there we remain throughout the entire story. This isn't to say we won't feel what other characters feel as well. But ultimately, what other characters do, think, and feel will *itself* be measured by its effect on the protagonist. It is the protagonist's story, after all, so we evaluate everyone and everything else based on how they affect him. Because ultimately what moves a story forward are the protagonist's actions, reactions, and decisions, rather than the external events that trigger them.

Your protagonist's reaction can come across in one of three ways:

1. **Externally:** Fred is late; Sue paces nervously, stubbing her toe. It hurts. She hops on one foot, swearing like a sailor, hoping she didn't chip the ruby red polish Fred loves so much.

2. **Via our intuition:** We know Sue's in love with Fred, so when we discover that the reason he's late is because he's with her BFF, Joan, we instantly feel Sue's upcoming pain, although at the moment she has no idea Fred even knows Joan.

3. **Via the protagonist's internal thought:** When Sue introduces Fred to Joan, she instantly senses something is going on between them. Watching them pretend to be strangers, Sue begins to plot the intricate details of their grisly demise.

When the events of the story are filtered through the protagonist's point of view—allowing us to watch as she makes sense of everything that happens to her—we are seeing it through her eyes. Thus it's not just that we *see* the things she sees—it's that we grasp what they *mean* to her. In other words, the reader must be aware of the protagonist's personal spin on everything that happens.

This is what gives narrative story its unique power. What sets prose apart from plays, movies, and life itself is that it provides direct access to the most alluring and otherwise inaccessible realm imaginable: someone else's mind. Lest the significance of this be lost, bear in mind that our brain evolved with just that goal—to see into the minds of others in order to intuit their motives, thoughts, and thus, true colors.[6] (We'll explore this further in chapter 4.) Even so, in life the key word is *intuit*; movies have the raw power to convey thoughts visually, through action; plays, via dialogue. While all three can be incredibly compelling (especially life), ultimately, they still leave us guessing. In prose, those thoughts, clearly stated, are where the story lives and breathes, because they directly reveal how the protagonist is affected by—and how she interprets the meaning of—what happens to her.

That's what readers come for. Their unspoken hardwired question is, *If something like this happens to me, what would it feel like? How should I best react?* Your protagonist might even be showing them how *not* to react, which is a pretty handy answer as well.

So how do you clue the reader into the protagonist's thoughts so we're privy to how he is making sense of what befalls him? That is, how do you let us know what he is *actually* thinking, especially since it's often the opposite of what he says? This is a doubly crucial question, since very often a character's reaction to what happens is solely internal—be it an unspoken monologue, a sudden insight, a recollection, or an epiphany. How you weave it into your story depends on whether you're writing in the first person or the third person. Let's take a quick look at each.

Conveying Thoughts in the First Person

Conveying the protagonist's thoughts when writing in the first person sounds like a no-brainer. After all, since he's telling us his story, *everything* reflects what he thinks, right? Exactly. This is what makes it tricky. Why? Because it means that every single thing in a story told in the first person must have a direct, implicit, and illuminating spin. Thus the narrator's opinion is laced into everything he tells you. Each detail he chooses to convey reflects his mindset and reveals something about him and how he sees the world. Think of it as the *Rashomon* effect: if four people witness the same event, you end up with four very different accounts of it—each one believable. Is one account true and the other three not? Nope, it's just that given each person's worldview, they processed what happened differently; each found certain aspects compelling, assigned them meaning as *each* saw it, and so drew different conclusions.

Is there an objective truth? Maybe. But considering that by definition we experience everything subjectively, how would we know? Which means that in a first-person account, everything the narrator tells us is imbued with his own subjective meaning, simply by virtue of the fact that *these* are the details he's picked to tell his story.

How is this different from the spin things have when writing in the third person? It's a difference in distance. In a third-person narrative, there are times when the *reader* evaluates the meaning of things relayed by the omniscient narrator (that's you, by the way), based on what he or she knows about the protagonist. For instance, the fact that Ted decided to surprise Ginger with a brand new plush Day-Glo orange couch is in and of itself neutral. But if we know that Ginger loved her old couch, hates orange, and don't even ask her about plush, then we'll have a pretty good idea how she'll feel when she sees it—regardless of what she says to Ted.

In a first-person account, on the other hand, nothing is ever neutral, even for a moment. This means the narrator will never tell us about anything that does not in some way affect him. He won't give us long objective passages about what the town looked like, what Edna wore to the office, how great the madeleine tasted, or how the Reagan administration ruined the country. Sure, he might tell us all these things but *only* because they have a specific effect on the story he's telling. It might help to think of the narrator as a narcissist (but in a good way). Everything in the story relates to him or else why would he be telling us about it?

Thus the narrator's thoughts are laced through everything he chooses to report, and he draws a conclusion about everything he mentions. But he doesn't stop there. He isn't the least bit shy about directly expressing exactly what he thinks about, well, everything. Of course, he could be completely wrong about everything he says—first-person narrators are often unreliable, and part of the reader's pleasure is figuring out what's really true.

The only thing a first-person narrator can't tell us is what someone else is thinking or feeling. So if Fred is talking about his breakup with Sue, he can't say, "When I told Sue I was in love with Joan, she felt as if she'd been gut punched." But what he can say is, "When I told Sue I was in love with Joan, the color drained from her face as if she'd been gut punched." Fred can infer or guess how Sue felt, but he can't come out and say it with certainty—unless, of course, Fred is the type of character who always assumes he knows exactly how everyone else is feeling, in which case we'll know that Fred's assertion about Sue feeling gut punched is meant to tell us something about *Fred*, rather than how Sue actually felt.

But what if Fred feels nothing? Because, you see, he's in denial. So naturally he doesn't react to Sue's rapidly escalating hints that she knows all about him and Joan. Right? Sounds like a real Catch-22, doesn't it? After all, if you *know* you're in denial, the cat's sort of out of the bag. So, when writing in the first person—or the third for that matter—how on earth can you convey all the things Fred is *not* thinking?

It goes without saying that the one thing you don't want to do is have Fred think nothing. Instead, the way to convey that he's in denial is to show how he interprets all those hints Sue's giving him. In other words, how does he rationalize them? Being in denial isn't as easy as it sounds. It's not a "blank" state; rather, it takes a good bit of work. When it comes to maintaining our coveted sense of well-being, each of us is a quintessential spin doctor.[7] This means that Fred will work overtime to make sense of things that, to the reader, clearly have an altogether different meaning than the one he's just assigned to them.

To sum up, when writing in the first person, it helps to keep these things in mind:

- Every word the narrator says must in some way reflect his point of view.

- The narrator never mentions anything that doesn't affect him in some way.

- The narrator draws a conclusion about everything he mentions.

- The narrator is never neutral; he always has an agenda.

- The narrator can never tell us what anyone else is thinking or feeling.

Conveying Thoughts in the Third Person

One of the beauties of writing in the first person is that you never have to worry whether the reader will know whose thoughts you're trying to convey. Every thought belongs to the narrator. But writing in the third person is another story, especially because there are several variations. First, here's a quick rundown of the three most frequently used:

1. **Third-person objective:** The story is told from an objective external standpoint, so the writer never takes us into the characters' minds at all, never tells us how they feel or what they think. Instead, as with film (long rambling voiceovers not withstanding), that information is implied solely based on how the characters behave. If you're writing in third-person

objective, you'll show us the protagonist's internal reactions through external cues: body language, clothes, where she goes, what she does, who she associates with, and of course, what she says.

2. **Third-person limited (aka third person close):** This is very much like writing in first person, in that you can tell us only what one person—almost always the protagonist—is thinking, feeling, and seeing. Thus the protagonist must be in every scene, and aware of everything that happens; the only real difference is that you're using "he" or "she" rather than "I." And as with first person, you can't tell us definitively what anyone but the protagonist thinks or feels unless that person pipes up and actually says it out loud.

3. **Third-person omniscient:** Here, the story is told by an all-seeing, all-knowing, objective and (traditionally) trustworthy narrator (you), who has the power to go into every character's mind and tell us what they are thinking and feeling, have done, and will ever do. The trick, of course, is to keep track of all of it. And to stay behind the curtain at all times. Even a fleeting glimpse of the puppet master completely ruins the illusion that there are no strings attached.

Okay, so how do you convey thoughts when writing in either third-person omniscient or limited? You do this by using something that, as far as the reader is concerned, is akin to telepathy. Good stories do it so well that we don't even notice they're doing it. In fact, I'd venture to say you've probably read hundreds of books written in the third person that clued you into what the characters were thinking so deftly that when it comes to figuring out exactly how they did it, you're *still* wondering whether you should italicize or put quotation marks around sentences that must then be clearly tagged as thoughts. The answer is, none of the above. No italics. No quotation marks. No tags.

Once you master the art of slipping your characters' thoughts onto the page, the reader will be able to automatically differentiate a character's inner thoughts from the narrator's voice. Readers intuitively expect the protagonist to have an opinion, whereas you, as narrator, don't even exist as far as they're concerned—that is, as long as you keep your opinions to yourself. The narrative voice is almost always neutral, meaning that as omniscient narrator, you're invisible and just reporting the facts. Your characters, on the other hand, are free to express their opinion on whatever they so desire. As long as the reader knows whose head we're in—that is, who the point-of-view character is—you rarely need a preamble at all. For instance, this is from Elizabeth George's *Careless in Red:*

> Alan said, "Kerra."
>
> She ignored him. She decided on jambalaya with dirty rice and green beans, along with bread pudding. It would take hours, and that was fine with her. Chicken, sausage, prawns, green peppers, clam juice . . . The list stretched on and on. She'd make enough for a week, she decided. The practice would be good, and they could all dip into it and reheat it in the microwave whenever they chose. And *weren't* microwaves marvelous? Hadn't they simplified life? God, wouldn't it be the answer to a young girl's prayers to have an appliance like a microwave into which *people* could be deposited as well? Not to heat them up, but just to make them different to what they were. Whom would she have shoved in first? she wondered. Her mother? Her father? Santo? Alan?[8]

Brilliant, isn't it, the way George uses a mundane event like deciding to make jambalaya as the leaping off point into the heart of a decidedly different matter? Notice how she does it by giving us a ride on Kerra's train of thought as it pulls out of grounded reality and heads into metaphor, where microwaving people sounds like a good idea. Notice, too, how we instinctively know that it's not George who is asking these questions but Kerra herself. In fact, both the "she decided" and the "she wondered" could be eliminated, and we'd still know.

Often, a character's thoughts help establish voice and tone, thus setting the mood—beginning on the very first page. Here is the second paragraph of Anita Shreve's *The Pilot's Wife*. All we know thus far is that the protagonist, Kathryn, has been awakened before dawn.

> The lit room alarmed her, the wrongness of it, like an emergency room at midnight. She thought in quick succession: Mattie. Then, Jack. Then, Neighbor. Then, Car accident. But Mattie was in bed, wasn't she? Kathryn had seen her to bed, had watched her walk down the hall and through a door, the door shutting with a firmness that was just short of a slam, enough to make a statement but not provoke a reprimand. And Jack—where was Jack? She scratched the sides of her head, raking out her sleep-flattened hair. Jack was—where? She tried to remember the schedule: London. Due home around lunchtime. She was certain. Or did she have it wrong and had he forgotten his keys again?[9]

Notice how every fact in that attention-grabbing paragraph has meaning that compounds in light of each new detail. In other words, it adds up. What emerges is a candid portrait of Kathryn, her family life, and how she processes information as she struggles to quell the growing suspicion that something is very wrong. It's not just her thoughts, which are very simple, but her thought pattern—staccato, ragged, confused—that drives the scene forward. The minimal tags Shreve gives us—"She thought in quick succession," "She tried to remember the schedule"—serve to highlight the thoughts themselves, establishing a style, and a voice, that is fresh, compelling, and hard to resist.

But is a tag really necessary? Do we need the author to tell us that we've slipped out of the narrative voice and into the character's head? Nope. In this snippet from Elmore Leonard's *Freaky Deaky*, there is no tag or signifier at all:

> Robin watched him drink his wine and refill the glass. Poor little guy, he needed a mommy. She reached out and touched his arm. "Mark?" Felt his muscle tighten and took that as a good sign.[10]

Is there any doubt that it's Robin, rather than Leonard, who sees Mark as a poor little guy in need of a mommy? Yet there are no quotation marks, no italics, no "she thought," "wondered," "realized," or "mused." There is nothing at all in the text that flags this as Robin's opinion. Why? Because none is needed. We get it. Just as we understand that it's Robin's *opinion* that Mark's muscle tightening is a good sign. As far as Leonard is concerned, she could be totally wrong—which is one of the things that keeps us reading. We want to find out.

Notice that, as when writing in first person, a character in third person can't make a definitive statement about how anyone else feels or what they're about to do. Just as in life, characters can only assume. And very often that assumption then tells us something about the character making it, as evinced by Selevan and his self-possessed goth granddaughter, Tammy, again from *Careless in Red*:

> She nodded thoughtfully, and he could tell from the expression on her face that she was about to twist his words and use them against him as she seemed only too expert at doing.[11]

George is not telling us that Tammy *is* going to twist Selevan's words. Rather, it's *Selevan* who is drawing that inference from her expression. From this we learn three things: that he's positive it will happen; that it might not; and, most revealing of all, that he feels she misunderstands just about everything he says. So since *Careless in Red* is written in third-person omniscient, if George wanted to make it clear that Tammy has, in fact, not misunderstood Selevan, could she reveal it in the next sentence by taking us into Tammy's head?

No, she couldn't, because that would be committing the sin known as "head hopping."

Head Hopping

No matter whose point of view you're writing in, you may be in only one head per scene. Thus, since George began the scene in Selevan's head, there she must stay. Why? Because switching POV in the middle of a scene is often so jarring it instantly breaks the flow. It looks something like this:

Ann paced, wondering when Jeff would snap out of it and tell her what happened. Had he finally told his wife Michelle about them? Why did he look so heartbroken? She wanted it to be a good thing, but try as she might, she couldn't come up with a single hopeful reason why he would sit hunched in the corner of the couch, staring at the frayed rug that, she realized, was in dire need of cleaning. When she could stand it no longer, she turned to him, "Jeff, what is it? What's wrong?"

She knows, Jeff thought. I can feel it. Sure, I told Michelle about her. Who knew she'd laugh and say, "Go ahead, run off with that loser. I bet she's the kind of woman who has a houseful of dirty rugs." I've been such a fool. But how do I tell Ann it's over? Maybe if I just sit here and stare at the rug, she'll figure it out. Women are pretty intuitive that way, aren't they?

When Jeff didn't answer, Ann's heart sank. It could mean only one thing: he'd told Michelle, and she'd mentioned that dirty rug thing again. She's been obsessed with it, ever since she opened that rug cleaning business back in March. God, Jeff is such a fool!

Disorienting, isn't it? So why do writers do it? Because they see it as the only way to convey information that is crucial to the scene. But is it? Not exactly. There is, in fact, a language that speaks louder than words. Let's have a listen, shall we?

Body Language

Imagine you're walking down the street; you turn a corner, and two blocks up you see a figure ambling away. Although from behind it could be anyone, you instantly recognize your best friend. How? By his gait.[12] Welcome to the world of body language.

Body language is the one language it's impossible to really lie in. As Steven Pinker says, "Intentions come from emotions, and emotions have evolved displays on the face and body. Unless you are a master of the Stanislavsky method, you will have trouble faking them; in fact, they probably evolved *because* they were hard to fake."[13] In other words, body language is the first thing we humans learned to decode, because even back in the Stone Age we knew that what a person grunts and what he really means might be two very different things.

The same is true of your protagonist. In a story, the goal is to show us how a character really feels—especially when there's a big discrepancy between what he *wants* to say and what he *can* say—through his body language. The most common mistake writers make is using body language to tell us something we already know. If we know Ann is sad, why would we need a paragraph describing what she looks like when she's crying? Rather, body language should tell us something we *don't* know. At its most effective, it tells us what's really going on inside the character's head. This is why body language works best when it's at odds with what's happening—either by telling us something that the character doesn't want known

Ann pretends to be completely calm but can't stop her right foot from nervously jittering.

. . . or by dashing a character's expectations:

Ann expects Jeff to be glad he's finally left Michelle; instead, he sits there, hunched, staring mournfully at the embarrassingly dirty rug.

We feel Ann's pain, because the author made sure we already knew what she expected—that Jeff would return grinning, with luggage. Instead, he's come back frowning, with baggage. Unless we're aware of *both* what Ann wants and what she then gets instead, all the body language in the world will be rendered mute. This sounds obvious, but you'd be surprised how often writers forget to let us know what a character *hopes* will happen, so that when it doesn't, we have no idea their expectations have been dashed.

With that in mind, let's revisit Ann and Jeff, this time using body language to convey the information:

Finally, when she could stand it no longer, Ann turned to him. "Jeff, what is it? What's wrong? Did you tell Michelle about us or not?"

Jeff said nothing, slouching further into the sagging couch, eyes downcast as Ann paced back and forth with such force that dust flew from the mangy rug. She saw him glance at her face, his eyes quickly darting away, and quickened her stride.

Ann's heart sank; she knew Michelle must have mentioned the dirty rug thing again. Why else would he just sit there, staring at the damn carpet, the coward. He was probably waiting for her to figure it out and send him packing. Jeff's such a fool, she thought. I'm better off without him.

In this version, even though we don't hear the exact details of what Michelle said to Jeff, Ann's insights go a long way toward filling us in on their dynamic. And see how clearly we understood what Jeff was thinking simply by reading his body language? Since we know what Ann wants (Jeff), and what she's realizing she's not going to get (Jeff), her body language told us what was going on in her head as she paced. Just as Jeff's body language reveals his reaction to hers. It is purely visual. It works because, story-wise, we know what it means, emotionally, to both of them. Otherwise, the scene would be opaque. Sure, we'd know something intense is going on between them, but we'd have no idea what.

But wait. Instead of all that, couldn't you just leap in and tell the reader how they're feeling? And while you're at it, give us the heads-up on who's right and who's being a little bit of a jerk? I mean, what if the reader gets it wrong?

And that brings us to another common pitfall: editorializing. It's what writers resort to when they don't trust the reader to get it.

Hey, You're Not the Boss of Me

Make us feel, and believe me, we'll know who's right and who's probably not. *Tell* us what to feel, on the other hand, and what we'll feel is bullied. That's why when you're checking to make sure you've conveyed how the action affects the protagonist, you want to resist the urge to jump in and take it a step further, by telling the reader what to think or feel about it as well. Editorializing is perfectly fine in a newspaper (remember those?) op-ed piece, where the whole *point* is to tell readers what you believe they should think or feel. In a story, telling readers what to feel not only annoys them but pushes them right out of the story. The reader's goal is to experience the story on her own terms, not to have it explained to her or be herded toward a specific hard-and-fast conclusion. This even applies to something as seemingly benign as exclamation points. They're almost always distracting! Really!! What's worse, they pull the reader's brain right out of the story by giving her an overt order rather than trusting the story to trigger her reaction all by itself.

So if you want us to think that John's a bad guy, show him doing bad things. It's just like in life: Imagine your coworker Vicky is telling you about her next-door neighbor, whom you've never met. "That John," she says. "He's such a jerk; he's the most self-centered, unscrupulous man I've ever met." Now, even though this may be an absolutely accurate assessment of John, because you don't know him, nor do you have any idea what Vicky's basing *her* assessment on, you have no way of knowing whether it's true. But you *have* been listening to Vicky rant on and on about how awful he is. And it sounded sort of bitter. So now you're wondering what she did to make John into such a meanie. Which, of course, is the exact opposite of her intent.

But if, instead of telling you how to feel about him, Vicky tells you that John steals from his grandmother, shoves past everyone on the train, and spits in his boss's coffee, you'll not only agree with her, but you might dislike John even more than she does.

Your job is not to judge your characters, no matter how despicable or wonderful they may be. Your job is to lay out what happens, as clearly and dispassionately as possible, show how it affects the protagonist, and then get the hell out of the way. The irony is, the less you tell us how to feel, the more likely we'll feel exactly what you want us to. We're putty in your hands as long as you let us think we're making up our own mind. That's why, as omniscient narrator, it's probably not a good idea to write this:

> "I don't think I can marry you Sam," said Emily, in that condescendingly bitchy way women who think they're better than men all seem to have.

Sure, if this were a first-person narrative told by a bitter protagonist named Sam, it would be spot-on. However, if this is the author talking, he is probably telling us a wee bit more about himself than he intends. Which—not to frighten you—is par for the course. As Johann Wolfgang van Goethe, who knew a thing or two about dancing with the devil, said: "Every author in some way portrays himself in his works, even if it be against his will."[14] Which means it might be time to reexamine that old stalwart: "Write what you know."

MYTH: Write What You Know

REALITY: Write What You Know *Emotionally*

If your protagonist is a trumpet-playing-neurosurgeon-turned-CIA-agent stationed in Antarctica, yes, you'd better know something about each of those things. But in the larger sense, "Write what you know" doesn't refer as much to facts as to what you know *emotionally,* which translates to your knowledge of what makes people tick.

Writing what you actually know, however, is a dangerous game given our natural propensity to tacitly assume that others have the same knowledge and beliefs that we do.[15] This tendency drives what communication scholars Chip and Dan Heath have dubbed "the Curse of Knowledge." They explain, "Once we know something, we find it hard to imagine what it was like not to know it. Our knowledge has 'cursed' us. And it becomes difficult for us to share our knowledge with others, because we can't readily re-create our listeners' state of mind."[16]

When writers unconsciously assume the readers' knowledge of—not to mention interest in—what the writers themselves are passionate about, their stories tend to be wildly uneven. On the one hand, the writer is so familiar with his subject that he glosses over things the reader is utterly clueless about. On the other, it's way too easy for the writer to get caught up in the minutiae of how things "really work" and lose sight of the story itself. This is something that, for some reason, lawyers seem particularly prone to. Over the years, I've read myriad manuscripts in which the story comes to a screeching halt while the writer outlines the legal ramifications of every single thing, as if the reader might sue, should some fine point of jurisprudence be overlooked.

Equally treacherous is the common misconception that just because something "really happened" it's believable (read: makes sense). That's why it's always helpful to have Mark Twain's pithy observation close at hand: "It's no wonder that truth is stranger than fiction. Fiction has to make sense."[17]

How do you make it make sense? By tapping into what you know about human nature and how people interact, and then consistently showing us the emotional and psychological "why" behind everything that happens. Do you have to hammer this out to the nth degree before you start writing? Of course not. As novelist Donald Windham so astutely says, "I disagree with the advice 'write about what you know.' Write about what you need to know, in an effort to understand."[18]

And speaking of understanding, here's a final word to the wise: the bigger the word, the less emotion it conveys. In fact, the less it tends to convey, period, beyond the vague notion that the author is showing off. This is something that both fledgling writers and winners of the National Book Award can easily forget. To best illustrate the point, here's said award winner Jonathan Franzen talking about a letter he received from a reader: "She began by listing thirty fancy words and phrases from my novel, words like 'diurnality' and 'antipodes,' phrases like 'electropointillist Santa Claus faces.' She then posed the dreadful question: 'Who is it that you are writing for? It surely could not be the average person who just enjoys a good read.'"[19]

We are all that average person, and the enjoyment we get isn't frivolous; it's rooted in our biology. It's what allows us to leave our real life behind and tap into how it would actually feel to experience life in someone else's shoes. Big words? They're pebbles in those shoes, ironically, distracting the reader from the very story they're meant to tell.

CHAPTER 3: CHECKPOINT

Does your protagonist react to everything that happens *and* in a way that your reader will instantly understand? Can we see the causal link between what happened and why she reacted the way she did? Are we aware of what her expectations are so we can tell whether or not they're being met? And, if she isn't in the scene in question, do we know how what happens *will* affect her?

If you're writing in the first person, is *everything* filtered through the narrator's point of view? Remember, in the first person, the narrator doesn't mention anything that doesn't relate to the story and that doesn't already have his personal spin stamped on it.

Have you left editorializing to the op-ed department? The more you have a message you want to convey, the more you have to trust your story to do it. The joy of reading is getting to make up your own mind about what a story's ultimate message is. The joy of writing is being stealthy enough to stack the deck so your reader will choose yours.

Do you use body language to tell us things we don't already know? Think of body language as a "tell," something that cues your reader into the fact that all is not as it seems.

4

WHAT DOES YOUR PROTAGONIST *REALLY* WANT?

COGNITIVE SECRET:

Everything we do is goal directed, and our biggest goal is figuring out everyone else's agenda, the better to achieve our own.

STORY SECRET:

A protagonist without a clear goal has nothing to figure out and nowhere to go.

What the human brain does best, what it seems built to do, is think socially.

—MICHAEL GAZZANIGA

BEFORE THERE WERE BOOKS, we read each other. We still do, every minute of every day. We instinctively know everyone has an agenda, and we want to be sure that agenda isn't to clobber us, either metaphorically or with a hammer. What we're hoping for is kindness, empathy, and maybe a nice big box of chocolates. So it's interesting to note that the term "agenda" often carries a negative connotation, implying something decidedly Machiavellian, as in duplicitous, manipulative, and cunning. Truth is, agenda is just another word for goal—making it completely neutral and utterly necessary to survival.

In fact, Steven Pinker defines intelligent life as "using knowledge of how things work to attain goals in the face of obstacles."[1] Almost sounds like the definition of story, doesn't it? It's interesting, too, that the most common obstacle in both life and story is figuring out what other people *really* mean. That's no doubt why, as neuroscientists have recently discovered, our brain comes equipped with something they believe might be akin to X-ray glasses: *mirror neurons.*

According to neuroscientist Marco Iacoboni, who pioneered the research, our mirror neurons fire when we watch someone do something *and* when we do the same thing ourselves. But it's not just that we register what it would feel like physically; our real goal is to *understand* the action.[2] As Michael Gazzaniga has noted, thanks to mirror neurons, "Not only do you understand someone is grabbing a candy bar, you understand she is going to eat it or put it in her purse or throw it out or, if you're lucky, hand it to you."[3]

Mirror neurons allow us to feel what others experience almost as if it were happening to us, the better to infer what "others *know* in order to explain their desires and intentions with real precision."[4] But here's the kicker. We don't just mirror other people. We mirror fictional characters too.

A recent study, in which subjects underwent functional magnetic resonance imaging (fMRI) of the brain while reading a short story, revealed that the areas of the brain that lit up when they *read* about an activity were identical to those that light up when they actually experience it. Yes, yes, I can see those of you who've read steamy novels nodding sagely and thinking, *Uh, you needed a brain scan to tell you that?*

Here's what Jeffrey M. Zacks, coauthor of the study, has to say about the physical effect a story has on us: "Psychologists and neuroscientists are increasingly coming to the conclusion that when we read a story and really understand it, we create a mental simulation of the events described by the story." But it goes much deeper than that. As lead author of the study Nicole Speer points out, the "findings demonstrate that reading is by no means a passive exercise. Rather, readers mentally simulate each new situation encountered in a narrative. Details about actions and sensation are captured from the text and integrated with personal knowledge from past experiences. These data are then run through mental simulations using brain regions that closely mirror those involved when people perform, imagine, or observe similar real-world activities."[5]

In short, when we read a story, we really do slip into the protagonist's skin, feeling what she feels, experiencing what she experiences. And what we feel is based, 100 percent, on one thing: her goal, which then defines how she evaluates everything the other characters do. If we don't know what she wants, we have no idea how, or why, what she does helps her achieve it. As Pinker is quick to point out, without a goal, everything is meaningless.[6]

It's a sobering thought, isn't it? So in this chapter our goal is to zero in on how to define your protagonist's goal, since it's what bestows meaning on everything that happens. We'll examine the difference between her internal and external goals, which are often at odds with each other; explore how both are driven by the core issue she's struggling with; and discover how to create external obstacles for her that add drama rather than stop the story cold.

Everyone Has a Goal

Mirror neurons allow us to walk a mile in the protagonist's shoes, which means he has to actually be going somewhere. The good news is that everyone—real, fictional, or somewhere in between—has a goal. Even those who want to remain exactly as they are and never change an iota have a goal—in fact, it's the biggest challenge of all. Staying the same in the face of the constant onslaught of perpetual change is no easy task, no matter how snuggly you strap yourself into your La-Z-Boy recliner, how firmly you close your eyes, how deeply you stick your fingers into your ears, and how loudly you hum.

The even better news is that what your protagonist wants dictates how she will react to everything that happens to her. None other than former president Dwight D. Eisenhower perfectly captures the essence of a successful story: "We succeed only as we identify in life, or in war, or in anything else a single overriding objective, and make all other considerations bend to that one objective."[7]

In a story, plot-wise, what all other considerations bend to is the protagonist's external goal. Sounds easy enough, until you add the fact that what her external goal bends to is her internal issue—the thing she struggles with that keeps her from easily achieving said goal without breaking a sweat. As we'll see throughout, this internal struggle is what the reader came for, whether he's conscious of it or not. The driving question is: what would it cost, emotionally, to achieve that goal?

Let me give you a quick down-and-dirty example. In the movie *Die Hard*, what's John McClane's goal? To stop pseudo-terrorists from murdering everyone at the company Christmas party at Nakatomi Plaza? To kill Hans Gruber? To live to see the dawn? Sure, he wants to do all those things. But his goal, which the movie makes clear in the very first scene, is to win back his estranged wife, Holly. And so everything that happens forces him to confront the reasons she left him and to overcome them, while at the same time running barefoot through broken glass, dodging machine gun fire, and leaping into fifty-story elevator shafts.

No Goal, No Yardstick

If you don't provide your protagonist with a driving deep-seated need that he believes his quest will fulfill, the things that happen will feel random; they won't add up to anything. Without knowing what he wants, or what his issue is, "There is no there, there," as Gertrude Stein so famously said (okay, she was talking about Oakland, California, but still). Without it, there's no yardstick by which to measure your pilgrim's progress, no context to give it meaning.

As a result, it's impossible to envision the coming chain of events—that is, the story itself. It's like watching football with no idea what the rules are, or how points are scored, or even that it's a game at all. Imagine that the protagonist, Hank, a massive man in a padded spandex uniform, catches a prolate spheroid (you wouldn't know it was a football). Suddenly, a whole bunch of other spandex-clad bruisers are rushing toward him. Now what? Should he run to the right, run to the left, throw it to the guy in the red uniform? Bury it, maybe? If you don't know what the objective is, everything appears random. The action doesn't add up, so there's nothing to follow, which makes it impossible to anticipate what will happen next. It is anticipation that creates the intoxicating sense of momentum that hooks a reader, so stories without it remain unread.

Making Meaningful Connections— Does It Add Up?

Before we dive deeper, it's important to keep one thing in mind. It's something we live by when we read but tend to forget as writers: readers assume that everything the writer tells them is there on a strictly need-to-know basis. Our assumption is that if we don't need to know it, the writer won't waste precious time telling us about it. We trust that each piece of information, each event, each observation, matters—right down to how the protagonist's hometown is described, the amount of hair gel he uses, and how scuffed his shoes are—and that it will have a story consequence or give us insight we need in order to grasp what's happening. If it turns out that it doesn't matter, we do one of two things: (1) we lose interest, or (2) we try to invent a consequence or meaning. This only postpones our loss of interest, which is then mingled with annoyance, because we've invested energy trying to figure out what the writer was getting at, when the truth is, she wasn't getting at anything.

But by figuring out what your protagonist wants, and the inner issue she'll have to overcome to get it, you can shape her quest with a confidence born of knowing you have a sturdy framework to guide you. For example: Wanda wants love, so her goal is to find the perfect boyfriend or, barring that, a nice golden retriever, preferably one who's good at fetch. This then becomes the story's single overriding objective and—you guessed it—the story question: will Wanda find love, human or otherwise? This is the info we're hunting for when we begin reading a novel. It's what tells us how the protagonist will react to what happens. So when Seth makes goo-goo eyes at Wanda, we'll know her heart is swelling, whereas if she wasn't so desperate for love, she'd surely see him for the sappy fool we know he is.

But of course there is a wee bit more to it than that. We still don't know what her inner issue is. Remember, it is the job of a story to dig beneath the surface and decipher life, not just to present it. Stories illuminate the meaning the protagonist reads into events that, in real life, would not be so easy to understand. Julian Barnes sums it up nicely: "Books say: *she did this because*. Life says: *she did this*. Books are where things are explained to you; life is where things aren't."[8]

In this case, what needs to be explained is *why* the protagonist wants what she wants, what it means to her, and what getting it will cost her. It's this that we, as readers, "try on for size." Cognitive psychology professor and novelist Keith Oatley puts it this way: "In literature we feel the pain of the downtrodden, the anguish of defeat, or the joy of victory, but in a safe space. . . . We can refine our human capacities of emotional understanding. We can hone our ability to feel with other people who, in ordinary life, might seem too foreign—or too threatening—to elicit our sympathies. Perhaps, then, when we return to our real lives, we can better understand why people act the way they do."[9] Or put more simply, as the aggravated newsreel producer barked at the beginning of *Citizen Kane*, "Nothing is ever better than finding out what makes people tick." Because with that comes the predictive power of knowing when to hold 'em, when to fold 'em, and when to run for cover.

And so, simply knowing that Wanda wants a boyfriend real bad isn't enough. We also have to know why and what issue she needs to come to grips with before she can succeed. Because there's no way she woke up one morning and bang! out of nowhere decided she can't live one more day without a mate. And don't try the old "but that's exactly the way it happened to my friend Susan" argument. Remember, a story can't get away with the things life can—and believe me, Susan actually *had* a good reason for it, whether she knew it or not. This is of key importance, so I want to pounce on it: *No one ever does anything for no reason, whether or not they're aware of the reason.* Nothing happens in a vacuum, or "just because"—especially in a story. The whole point of a story is to explore this "why" and the underlying issue that, in real life, dear old Susan never let on she was struggling with. Otherwise, how will we, as readers, be able to pick up pointers for navigating our own lives?

Thus the protagonist's true goal—even if it's *triggered* by a random external event—is something that's been evolving for years, although he might have been completely unaware of it until that very moment. That's because his desire stems from what it means to him internally, rather than merely what it does for him externally. For instance, Norm doesn't want that million dollars just because of all the shiny things he can rush out and buy the second he gets it. He wants it because all his life he's believed having a lot of money is what proves you're a real man, although of course this is not something he'd admit to anyone, including himself. It is, however, what drives his action. It helps to think of the source of your protagonist's desire as the answer to the question actors always ask: *What's my motivation?* Because as we know, the heart of the story doesn't lie in what happens; it beats in what those events mean to the protagonist.

Case Study: *It's a Wonderful Life*

Let's take a look at a film that's much beloved—and hard to avoid, even for the curmudgeons among us: *It's a Wonderful Life.* It's pretty clear from the start that protagonist George Bailey's goal is to get the hell out of Bedford Falls. Why? Because, as he tells his father, the thought of being chained to a rickety desk for the rest of his life would kill him. He wants to do something that matters, something big that people will remember. In short, George equates staying in Bedford Falls with being a failure, which means if he stays there, no matter what happens, he couldn't possibly be a success—this is the inner issue he's battling. And it gives him a pretty powerful motivation for getting the hell out. This sentiment underlies everything he does. It's what he struggles with each time something threatens his getaway.

By the same token, what keeps George in Bedford Falls aren't the external events that befall him, either—it's not his father's death, it's not that his brother Harry doesn't really want to take over Bailey Building and Loan, it's not the run on the bank. What stops George from leaving is also internal: his integrity. He can't leave because, as much as he wants to, he knows people are counting on him. Thus what fuels his external reaction to these events is his *internal* struggle. It is what causes him to make the choices he does. Notice, too, that all this revolves around what we've learned from neuroscience: the brain is built to think socially. It's not what happens externally that motivates George; it's the responsibility he feels toward others, and how he sees himself.

George's greatest reward is, of course, internal as well. That's why it doesn't matter that no one except warped, frustrated old man Potter ever *really* knows what happened to the missing eight thousand dollars, or that no one ever proves George didn't embezzle it. Think about it: when the movie ends, for all anyone knows, George may actually have stolen it and buried it out in Bailey Park. The point is, it makes no difference, because being vindicated on the plot level is small potatoes compared to George's real reward: the *internal* knowledge that all the concessions he made didn't rob him of the life he wanted—as a matter of fact, on reflection, George realizes they *gave* it to him. What's more, George's epiphany occurs *before* everyone shows up at his door, ready to bail him out. If they'd carted him off to jail that night, he'd have gone a happy man.

But they didn't, because the other characters responded in kind; their true gift to George is internal, as well. Sure, on the plot level they give him the money to stay out of jail. But what they *really* give him is unconditional love—as hokey as that sounds. He spent his life doing what integrity demanded. And that's exactly what everyone in Bedford Falls does when they believe George is down for the count. As Uncle Billy tells him, when Mary let people know he was in a bind, no one asked what had happened; they were too busy reaching into their pockets and asking what they could do to help.

Proust observed, "The only true voyage of discovery . . . would be not to visit strange lands but to possess [new] eyes."[10] That is exactly what happens to George Bailey: he looks back on his life with new eyes and sees something altogether different from what he expected. And in so doing, he makes a discovery often made by protagonists: his external goal and his internal goal were at odds all along.

Upon Achieving the Internal Goal, Revisiting the External Goal

Often the protagonist's external goal changes as the story progresses—in fact, that's often what the reader is rooting for (remember Scarlett?). In *It's a Wonderful Life*, George's internal goal is to make a big difference in the world. His external goal is to get out of Bedford Falls and build bridges, skyscrapers, and to "do big things." He believes these goals are one and the same. The movie then chronicles how his external goal is thwarted at every turn, and instead of doing big things, he always does the right thing. In the end, that's precisely how he achieves his internal goal—making a big difference in many people's lives—which brings with it the realization that he actually achieved his external goal as well. He *did* do big things—things that are far more important and enduring than building skyscrapers. Thus, by achieving his internal goal, he was able to redefine his external goal—and, happily, discover that he'd already accomplished it.

But up to that moment, George fully believed that only by achieving his external goal would his internal goal be met. And as real life makes all too clear, this is rarely the case. How many of us have thought, if only I could lose ten pounds (external goal), my life would be perfect and I'd be happy (internal goal)? Fueled by the belief it's a twofer—achieve the external goal and the internal goal will follow—we lose those ten pounds (and the hard way, no less, without lap belts, stomach stapling, or liposuction). That's when we discover—alas!—our lives are still not perfect, and now we're even *less* happy because at least when we were fat we could fantasize about how great it would be once we were thin. It's only then that we see the fallacy of our original assumption and begin wondering what, exactly, we really do need in order to be happy. By defining your protagonist's internal and external goals, and then pitting them against each other, you can often ignite the kind of external tension and internal conflict capable of driving an entire narrative.

The Real Issue: The Protagonist as Her Own Worst Enemy

What the protagonist must overcome to achieve her external goal tends to be pretty straightforward—that is, the external plot-driven obstacles that stand between her and success—but what about her internal goal? What stands in the way of that? In the you-have-to-fight-fire-with-fire category, the answer is, internal obstacles—usually in the form of longstanding emotional and psychological barriers—that are forever holding her back. This, then, is her internal issue. We're talking about the fear that whispers, *What the hell do you think you're doing?* as she approaches each hurdle. It's a voice that gets more convincing as the hurdles escalate in difficulty until by the end, the protagonist stops dead, sure there is absolutely no way she'll be able to overcome that last hurdle—not with that voice nattering in her head, anyway. While in real life such a person might just pop a Prozac and watch the problem recede into a comforting haze, in a story she has to do it the old-fashioned way—cold sober and on her own.

In order to construct these internal obstacles, ask yourself: *Why* is the protagonist scared? What, *specifically*, is she scared of that keeps her from achieving her goal? By now I'm guessing you know that the answer probably isn't, *She's afraid of losing her true love, going broke, or dying.* Even though, plot-wise, that's *exactly* what she's afraid of. Hell, it's what we're *all* afraid of—hence it's general and generic, and it doesn't tell us anything we don't already know. So although it's a good beginning, it is *only* the beginning.

Like the protagonist's goal, her fears spring from, and are defined by, her life experience—something we're going to be talking about in depth in chapter 5. But for now, let's take the most obvious fear: fear of death. I don't blame you if you're thinking, *Oh come on—that needs an explanation*? It's universal, you don't have to "learn" anything to know that the last thing you want on your daily to-do list is *Toddle off into the great beyond.*

Fair enough. I'm not going to argue with that. I'm going to sidestep it, because it's not the issue. The issue is: what does dying, *at this minute*, mean to the protagonist? For instance, who will she leave behind who needs her now more than ever? What won't she accomplish that she swore on her mother's grave she would? What burning promise won't she be able to keep? What wrong must she live till dawn to set right? The answers to these questions will tell you what dying means to your protagonist, beyond the big "Uh-oh!"

Yep, it always comes back to this: what do these events *mean* to the protagonist? What *is* her true goal? Knowing this will allow you to make her goal specific to her, rather than leaving it as a surface (read: generic) goal that we all have.

Why, then, do writers lob generic problems at their protagonists all the time? Sadly, it's often because they're following one of the great myths of storytelling.

MYTH: Adding External Problems Inherently Adds Drama to a Story

---

REALITY: Adding External Problems Adds Drama
***Only* If They're Something the Protagonist Must**
Confront to Overcome *Her Issue*

The myth that external problems add drama has plagued writers from time immemorial and has been inadvertently perpetuated by the myriad versions of the "hero's journey" story-structure model, which mandates that certain external events must happen at certain specific points in a story. The result is that writers craft *plots* in which these events occur rather than crafting *protagonists* whose internal progress depends on said events occurring. Such stories are written from the outside in: the writers throw dramatic obstacles in their protagonist's path because the timeline tells them to rather than because they're part of an organic, escalating scenario that forces the protagonist to confront her inner issue. Thus the dramatic events aren't spawned by the story itself but by an external by-the-numbers story-structure formula.

To create organic, compelling obstacles that work, you must make sure that everything your protagonist faces—beginning on page one—springs specifically from the problem she needs to solve, both internally and externally. This will help you avoid a very common pitfall: using a generic "bad situation" to create the protagonist's goal.

I've read countless manuscripts that began encouragingly in the midst of an upheaval: the protagonist's husband just walked out; the protagonist is driving to work when a huge earthquake strikes; the protagonist missed the jetty back to the cruise ship and now she's stranded in Venezuela with nothing but what she's wearing—a string bikini and flip flops. This is all good. The problem was, these authors had merely plunked their protagonist into a dicey situation to see what would happen next. But because the protagonist didn't have a long-standing need that was then put to the test, her "goal" was nothing more than getting out of the horrible position she unexpectedly found herself in. Thus the spotlight remained on the *problem* rather than the *protagonist*. Sure, things happened, but they didn't affect the protagonist on anything but a surface level. Because we had no clue what her specific desires, fears, or needs were beyond the very obvious one-dimensional need to get out of the current situation ASAP, we couldn't anticipate how she would react to the things that happened, except in a generic, that's-what-any-person-would-do sense. And that, my friends, is boring. Why? Because we all have a pretty good idea of what "any person" would do. Where's the suspense in that? We turn to story to tell us something we don't know. So while we don't care a whit about what "any person" would do, we care passionately about what your protagonist would do—as long as we know why.

Having a firm understanding of what your protagonist's specific goals and fears mean to her provides you with concrete plot guidelines. For instance, let's take the disappointing manuscript that opened with the protagonist's husband walking out. This was the story: the wife, Deb, blindsided by her husband Rick's unexpected departure, simply picked herself up and got on with her life, rather than whining about it (which was too bad, because a little pointed whining would have at least given us *some* clue as to what their marriage was like, who she is as a person, and what her personal arc might be). Trouble was, with no real problem predating the breakup of her marriage, Deb was way too well adjusted to be interesting—so well adjusted, in fact, that the reader immediately wondered both why Rick left her and why she had married such a deadbeat in the first place. Ironically, that was the *only* sign that there might be more to Deb than met the eye, but since it was never developed, it read as what it was: a plot convenience.

So, does Deb's story need to be scrapped? Not necessarily. Let's take a shot at developing Deb's dilemma ourselves, shall we?

THE STORY OF DEB'S BAD MARRIAGE

First stop, Deb's backstory (something we'll be talking about in far greater depth in the next chapter). What if Deb had stayed in a bad marriage because she didn't have the courage to admit, even to herself, that she was terrified she couldn't make it on her own? Thus Deb's goal isn't simply to move past a bad situation; it's to overcome a problem that *preceded* (if not caused) her current dilemma. Now we've expanded the premise: when Deb's husband walks out on her, she's forced to see whether she can, indeed, make it completely on her own—the one thing she's always feared most. This is a much bigger and more engaging question, one that opens the door to a whole slew of follow-up questions worth exploration:

- What caused Deb's fear that she couldn't be self-sufficient?

- Did that fear cause her to marry Rick in the first place?

- Was she settling?

- Was entering into a bad marriage her way of avoiding having to prove herself?

- Did Deb's fear perhaps make her a tad passive-aggressive, and so Rick's bad behavior wasn't as one-sided as it appears at first blush?

- In fact, was dealing with the daily drama of her failing marriage what actually kept her in it, because it diverted her from having to come to grips with her biggest fear?

Wouldn't you read on to find out?

But wait—now that we've mapped out the roots of Deb's goal and fears, how *do* we get them onto page one without beginning: "Deb was born in 1967 in a little cottage. . . ." Remember, we're not trying to tell the reader everything there is to know about Deb and her predicament on the first page, we're simply trying to *imply* that there is a lot to know. Our goal is to make the reader *feel* like they know her, and—this is essential—to care enough about her to want to find out what will happen to her. Which means we've also got to establish two things—that big changes are coming and all is not as it seems—and we have to do it as quickly as possible. Let's give it a try:

Shifting the weight of the grocery bags, Deb slid the key into the lock and braced herself. Not that Rick ever hit her—something that bad, and she'd actually leave. It was six, so she knew he'd be home. The TV would be on. And he'd ignore her with such intensity it would be like walking into a headwind. She told herself she hated him, angry that her pulse quickened anyway. It had been another dull day. Shopping, cleaning, exercising as if it mattered. It struck her this was the first time since Rick's sullen departure for work that morning that she'd been aware of her senses at all. There was the sound of a car pulling out of a driveway. The smell of the leaves that had been moldering under a tarp in a corner of the front yard since fall. With a sigh she turned the key, felt the click in her fingertips. The door swung open and she stumbled into the silence.

The house was empty. No Rick. No furniture. Nothing but a plain white envelope propped on the mantel, with her name neatly typed across it.

Can you see the elements of Deb's backstory planted there? For instance, "Not that Rick ever hit her; something that bad, and she'd actually leave," tells us that, in Deb's view, Rick *has* been doing bad things to her, but that, short of hitting her, she considers it livable (suggesting that Deb is an ace rationalizer). The phrase, "Shopping, cleaning, and exercising as if it mattered," tells us that being in shape hasn't netted her much—perhaps Rick hasn't noticed? It's pretty obvious what's meant by the sentence, "She told herself she hated him, angry that her pulse quickened anyway"—although there is enough ambiguity here to make us wonder about it. This sentiment is then echoed in: "It struck her this was the first time since Rick's sullen departure for work that morning that she'd been aware of her senses at all," which also gives us a glimpse of what Rick is like—at least according to Deb. Next, the example of what Deb then hears and smells aren't random just-because-they-were-there sensory details, but each has a definite subtext: "the sound of a car pulling out of a driveway" (we're about to find out that Rick has left her; maybe it was him in that car?); "leaves that had been moldering under a tarp in a corner of the front lawn since fall" (just as Rick and Deb's marriage was allowed to decay in plain view). And finally, as we discussed in chapter 3, notice that although the story is written in third person, it's clear we're in Deb's head, viewing everything from her point of view.

From there Deb's story seems pretty clear-cut: it's the tale of a conflicted woman whose husband has lost interest in her and is probably heading out for greener pastures. Or is he? Because so far, all we have is her side. What about Rick's? Could part of what Deb needs to overcome be a fundamental misunderstanding of what Rick's agenda actually is?

Case in Point: *The Threadbare Heart*

The foundation of a story is often rooted in just this kind of misinterpretation, which arises from the fact that because Deb isn't a literal mind reader, she interprets what her mirror neurons tell her based on her own understanding of the world and what *she'd* mean if she were Rick. We all do it. Someone does or says something that sounds hurtful, and we're hurt. But sometimes that hurtful utterance, which can trigger a story's arc, turns out to mean the exact opposite of what the protagonist thought it did.

Jennie Nash's keenly insightful novel, *The Threadbare Heart*, turns on just such a natural misunderstanding. The protagonist, Lily, has been married to Tom for over twenty-five years. It's been a good marriage, and Lily believes she knows Tom deeply and that their bond is solid. On page five, however, feeling safe, secure, and happy, she decides to risk having a bit of chocolate, knowing it might trigger one of her debilitating migraines. Noticing, Tom demands to know what she's doing, and she tells him not to worry: she'll deal with the consequences should she get a headache. She's stunned when he then angrily informs her in no uncertain terms that her headaches are *his* problem and always have been, at which point he stomps out. Suddenly, she's not sure if she knows him as well as she thought she did, and the world feels like a far more dangerous place. The reader, too, immediately shares Lily's unease—for about four pages. And then on page nine, Tom reflects on what happened:

> Lily's headaches were something he had handled for years without complaint. But the last few times, they had grabbed hold of him in a way that frightened him. He had imagined Lily spiraling further down into pain than she had ever gone before, spiraling so far away that she was out of reach. It made him think about her dying and his being alone. That wasn't something he felt like he could endure.[11]

This is a clear case of the "why" changing the surface meaning of an event by 180 degrees (which is exactly the sort of info readers are hungry for). What motivated Tom's outburst is the opposite of what, at first blush, it appeared to be. It wasn't that Tom was angry at Lily for risking a headache; it was that he loved her so much, he couldn't bear anything—including her own pain—that might take her away from him. Ironically, Lily *does* know Tom as well as she thought she did, but now she's not so sure. She no longer knows how much her husband loves her, but we do. And so as she struggles with it over the course of the novel, we're able to gauge her progress against what we know are Tom's true feelings for her. This is possible only because we are aware of not only Lily's agenda, but of Tom's as well.

It is this kind of glimpse into someone else's hopes and fears that makes stories so compelling—and so much more than mere entertainment. It's *hard* to understand what other people want from us. It's *hard* to know what we truly want for ourselves (well, besides another piece of that salted caramel chocolate). Stories not only give us much-needed practice in figuring out what makes people tick, they give us insight into how we tick.

CHAPTER 4: CHECKPOINT

Do you know what your protagonist wants? What does she desire most? What is her agenda, her *raison d'être*?

Do you know *why* your protagonist wants what he wants? What does achieving his goal mean to him, specifically? Do you know why? In short, what's his motivation?

Do you know what your protagonist's external goal is? What specific goal does his desire catapult him toward? Beware of simply shoving him into a generic "bad situation" just to see what he will do. Remember, achieving his goal must fulfill a longstanding need or desire—and force him to face a deep-seated fear in the process.

Do you know what your protagonist's *internal* goal is? One way of arriving at this is to ask yourself, *What does achieving her external goal mean to her?* How does she think it will affect how she sees herself? What does she think it will say about her? Is she right? Or is her internal goal at odds with her external goal?

Does your protagonist's goal force her to face a specific long-standing problem or fear? What secret terror must she face to get there? What deeply held belief will she have to question? What has she spent her whole life avoiding that she now must either look straight in the eye or wave the white flag of defeat?

5

DIGGING UP YOUR PROTAGONIST'S INNER ISSUE

COGNITIVE SECRET:

We see the world not as it is, but as we believe it to be.

STORY SECRET:

*You must know precisely when, and why,
your protagonist's worldview was knocked out of alignment.*

My life has been full of terrible misfortunes,
most of which never happened.

—MICHEL DE MONTAIGNE

WHEN I WAS FIVE, I closed my eyes and thought really hard about whether I was now invisible. After all, I couldn't see anything, so how could anyone see me? *Yes*, I concluded: I had indeed vanished. It made perfect sense, and it was thrilling to boot. I felt very smart. And why not? Being wrong feels exactly like being right, as journalist and self-proclaimed "wrongologist" Kathryn Schulz so brilliantly points out in her book *Being Wrong*. I spent days plotting how to walk blind-folded through the kitchen without bumping into anything, the better to secretly borrow a few cookies. That is, until I heard my mother asking what the heck I was doing with my hand in the cookie jar—which opened my eyes, both literally and figuratively.

Being wrong changes how we see—or don't see—the world. And we're wrong a lot, partly because in order to survive, we're wired to draw conclusions about everything we see, whether or not we have all—or any—of the facts; and partly because, more often than not, it's our cognitive unconscious that deftly constructs the implicit beliefs that then organize and rule our world.[1] So while it may be cold comfort, being wrong usually isn't our fault, at least not in the "you know damn well what you did" sense. Most times, we have no clue. According to neuropsychologist Justin Barrett, our implicit or "nonreflective" beliefs are our default mode, constantly working behind the scenes to shape memory and experience.[2]

As a result, from the moment one of those erroneous implicit beliefs is formed—*everyone's only in it for themselves, so the nicer*

85

someone is, the more you know they're out to con you—we blithely misinterpret everything that happens to us. *Everyone here is so nice— I better watch my back.* And the scary thing is, we don't even know we're doing it until something happens that proves us wrong, and suddenly our implicit belief is catapulted into our conscious mind, where we have to either deal with it or work overtime to rationalize it away.[3]

Stories often begin at just that moment, as one of the protagonist's long-held beliefs is about to be called into question. Sometimes that belief is what stands between her and something she really wants. Sometimes it's what's keeping her from doing the right thing. Sometimes it's what she has to confront to get out of a bad situation before it's too late. But make no mistake, it's her struggle with this "internal issue" that drives the story forward. In fact, the plot itself is cleverly constructed to systematically back her into a corner where she has no choice but to face it or fold up her tent and go home. The events relentlessly cajole and coax her to reexamine her past, which often looks— and feels—very different in retrospect. It's the same way that in life the present continually prods us to reassess our autobiographical selves, and as a result, past "events acquire new emotional weights . . . [and] facts acquire new significance."[4] Or as T. S. Eliot so aptly noted, "The end of our exploring will be to arrive at where we started, and to know the place for the first time."[5]

Which brings us to a trick question: when you're writing a story, where *is* the best place to start? No, the answer isn't *at the beginning, on page one,* or even, *sitting at my desk.* The best place to start working on a story is long before your poor unsuspecting protagonist shows up on page one. The best place to start is by pinpointing the moment long before, when she first fell prey to the inner issue that's been skewing her worldview ever since.

That's why in this chapter, we'll tackle something writers often shy away from: the notion of getting to know their characters before they tell their story. To that end we'll examine the very important pros and the trivial cons of outlining (how's that for a great example of editorializing?); why it's important to write focused character biographies that, happily, often beget outlines on their own; and why an exhaustive character bio can be more damaging than not writing one at all. And then, lest we get lost in the conceptual, we'll run though an example of just how it's done.

You Can't Fix It If It Ain't Broke

Stories are about people dealing with problems they can't avoid—sounds so elementary, doesn't it? Why, then, do writers so often leap in without knowing what, exactly, the protagonist's problem actually is? Often it's because they're hoping it'll become clear if they just start writing. But if you don't know what's broke, how can you write a story about fixing it? Which is why the second most-frequent editorial note that writers get, right after, "Uh, what's this story about, anyway?" is "Why now?" as in, why does the story start at *this* minute as opposed to yesterday, tomorrow, or when Aunt Bertha gets back from bingo?

Ironically, often the same writer who swears that it would crush her creativity to pause to outline or work out character bios will start the story at the exact spot in the protagonist's past where, instead, she should be digging; that is, at the moment his worldview was knocked out of alignment, along with the inception of the desire it thus thwarts. What she doesn't realize is that the story itself actually begins much later, when those two long-dormant opposites come to a head, giving the protagonist no choice but to take action. This concept is elegantly summed up by, ahem, the Oracle to Optimus Prime in the animated TV show *The Transformers: Beast Machines*: "The seeds of the future lie buried in the past."[6]

Does this mean you really have to outline your story first? Sure sounds like it. But like everything else, it's relative. Let's take a look at the arguments for and against.

The Great Outlining Debate

Many very successful authors swear the only way they can write is to jump in cold on page one, armed with nothing but the vaguest notion of where they're going. For them, the kick is to uncover the story as they write it. If they've already figured it out, the thrill is gone and the actual writing feels redundant.

For instance, there is the legendary (read: probably apocryphal) story of Edith Wharton who, after a manuscript she'd just completed was lost in a fire, told her editor that she couldn't possibly rewrite it, because she already knew the ending. In this Robert Frost concurs: "No surprise in the writer, no surprise in the reader."[7] Ditto Robert B. Parker, who says he has no idea where the story is going when he starts writing.[8]

And then there's the other school, with members like Katherine Anne Porter, whose philosophy is Ms. Wharton's polar opposite: "If I didn't know the ending of a story, I wouldn't begin."[9] Or how about none other than J. K. Rowling, who had very carefully plotted all seven Harry Potter books by 1992—when she began writing the first one.[10] "I spent an awful lot of time thinking about the details of the world and working it out in depth," she says. "I always have a base plot outline."[11]

Is either camp right? Or does this simply illustrate that it's up to each writer to decide whether outlining fits into his or her writing process and leave it at that? Probably. Then again, there's another way to look at it. Some lucky pups are simply born with a natural sense of story, the way some people have perfect pitch. They can toss off a laundry list and it comes out so nuanced and moving that you're weeping over the plight of poorly sorted socks. If you're one of those writers, you don't need me. Go forth and prosper! But most writers—including most successful writers—benefit from puttering around in their protagonist's past before tackling (or, having learned their lesson, rewriting) page one. Especially because it helps avoid two major pitfalls:

1. The most common problem with stories that haven't been outlined is that they don't build. How can they? Without a premeditated destination based on the battle between the protagonist's inner issue and his longstanding desire, they wander, taking the scenic route to who-knows-where. Thus, when the writer begins revising, something seminal needs to happen on, oh, about page *two*. And once it does, *everything* that follows becomes largely irrelevant. Which basically translates to what's known as a "page-one rewrite"—think of it as pretty much starting from scratch.

2. *Hey*, many writers think, *no biggie. I expect to rewrite. Everyone says that's a huge part of the process anyway.* Very true. But in this case there's a much bigger problem. It's extremely difficult to acknowledge that the first draft has been rendered largely moot. It's one of those hard-to-admit mistakes we were talking about, the kind we tend to work overtime rationalizing down to size. Thus new material is crafted first and foremost with an eye toward how it will fit into what's already there, because our unconscious allegiance is to what we've already written, rather than to the story itself. Ironically, the "new" draft is often a big step backward—what was flat in the prior version remains flat, now it just makes less sense.

Have I convinced you to give outlining a chance? Good. But before visions of rigid Roman numeral outlines fill your head—or worse, the thought of plowing through one of those endless one-size-fits-all "character questionnaires"—let me reassure you that outlining can be an intuitive, creative, and inspiring process. Not to mention one that's often surprisingly shorter than you might think. Let's take a look at why.

**MYTH: You Can Get to Know Your Characters
Only by Writing Complete Bios**

———————

**REALITY: Character Bios Should Concentrate Solely
on Information Relevant to Your Story**

When it comes to getting to know your characters, there is definitely such a thing as Too Much Information. We're not talking about details that are way too personal. In a story, way too personal is a good thing. But irrelevant is not. Yet writers are often told that in order to really get to know their characters, they must fill out a detailed character questionnaire longer than the book itself, answering questions like these (and for the record, I'm not making these up):

- Does he like his middle name?

- If she's stretching out in her backyard to sun herself, what kind of towel does she lie on?

- Does she have a favorite room?

- What color evokes strong memories for her?

- Does he have a birthmark?

- Does he have matching china?

- If he has a birthmark, is it by any chance in the shape of China? (Okay, I made that one up.)

- What is his opinion on euthanasia?

Now while the answers to all these questions might indeed be interesting, chances are they won't have anything to do with your story. The same goes for writing a general from-birth-to-the-present bio. The whole point of a story is to filter out the kind of unnecessary information such bios are full of. The trouble is, long character bios tend to be so all-encompassing that, ironically, they obscure the very info you're looking for. Here's the secret: you are looking *only* for information that pertains to the story you're telling. If a story is about a problem, then what you're looking for is the root of the problem that will blossom on page one. This means that if the fact that Betty is a virtuoso harpist doesn't enter into or affect the story, you don't need to make note of the grueling years she spent mastering the harp. Because if you do, you're likely to then waste time agonizing over where this fact should pop up in the story (when the truth is, it shouldn't), or worse, writing an entire subplot so Betty can show off her harp-playing skills at a holiday office party, which, because it has nothing whatsoever to do with the story you're telling, stops it cold. To add insult to injury, the harp problem doesn't end there; it lingers in the reader's mind, leaving him wondering, *Gee, I wonder where that harp playing thing is leading.*

That's why, when writing your protagonist's bio, the goal is to pinpoint two things: the event in his past that knocked his worldview out of alignment, triggering the internal issue that keeps him from achieving his goal; and the inception of his desire for the goal itself. Sometimes they're one and the same. For instance, in *It's a Wonderful Life,* this telltale moment is when George watches his father being beaten down by Potter. This leads George to believe that he can't be a success if he stays in Bedford Falls (skewing his worldview) and spurs him to want to be the success his father wasn't, by building big things somewhere else. The story then forces him to reassess his worldview until he slowly realizes that it—and his external goal—have been way off the mark.

While in many stories we wouldn't actually see this "telltale" scene, it's often referenced while the protagonist struggles with the havoc it wreaks on his life. It may not even be mentioned at all, its presence merely implied by his actions. So, although the reader doesn't see it, they feel its effect, because you, the writer, understood it so clearly that you were able to weave it through everything the protagonist does.

Thus when you write your protagonist's bio, the goal is to find those seminal moments and then trace the trajectory of events they triggered, culminating in the particular dilemma your story will revolve around. Once you've done that, if you're still dying to write an in-depth tell-all bio for your protagonist, who am I to stop you? But be forewarned: if you're not careful, some of those juicy-but-irrelevant details you've uncovered might then creep into your story on their own. The good news is that by using the techniques in this book, you'll be able to weed them out before they choke the life out of your story.

Then again, you just might find that once you've written your focused character bios, you're aching to dive into the story itself. With that goal in mind, when on the hunt for the telltale moments buried in your protagonist's past, it helps to remember these four do's and don'ts.

Do's and Don'ts for Writing Character Bios

1. **Do keep in mind one utterly-obvious-when-you-say-it-but-otherwise-easy-to-forget truism: story is about something that is *changing.*** Things start out one way and end up another—this is what is meant by a story's arc. The story itself unfolds in the space between the "before" and the "after." It chronicles the exhilarating time when things are in flux, giving the reader the illusion that it really could go either way. Thus what you're looking for when you write your character bios is the specific *before* that leads to the moment when suddenly everything is in flux. This *before* will yield information you'll then seed into the story you're telling so the reader understands what the protagonist is changing *from.* Look at it this way—a butterfly may be beautiful in and of itself, but what makes it *interesting* is that it used to be a caterpillar. The "before" is the yardstick that allows the reader to measure the protagonist's progress toward "after."

2. **Don't be uncomfortable about digging deep into your characters' psyches.** Don't hold back on account of decorum. You have an idea, going in, what their issues are—it's what you're writing about. Ask them embarrassing questions about it—the more personal, the better. Seek out the good, the bad, and especially the ugly, the messy, and the secrets they'd really rather keep to themselves. Nothing can be off limits. Rather than overlook their flaws, you want to pinpoint each one and then examine it under a high-powered microscope in light of their internal issue and their goal. *Your* goal is to allow them to be full, complete flesh-and-blood characters who, like us, are doing their best to muddle through against all odds. The essence of a story lies in revealing the things that in real life we don't say out loud. This is why, as cruel as it may feel, you can't allow your characters any privacy or mercy when exploring their past. Sure, they may demand it anyway; they may hide things from you; they may even lie to you. But if you *let* them hold back, if you let them hedge, the resulting story won't have the ring of truth. And don't kid yourself; the reader will know. We've been around the block, and since we automatically use our own default knowledge base to understand others—whether real or fictional—we have a pretty good idea where you are leading us when we start reading your story.[12] Hell, it's why we signed on in the first place. Veer away and we'll know it, lose interest, and go see what's on TV.

3. **Don't try to write well.** The good news about writing character bios is you can do it in a linear, straightforward—plodding, even—progression. Or you can jump all over the place if you like. It's your call entirely. Plus, there's no need to worry whether the first line hooks anyone or there are too many adjectives hanging around or even whether it's well written. All you're interested in is content; how it's presented is completely irrelevant, which,

ironically, often leads to stellar writing. Probably because it temporarily disconnects the often snide, hypercritical editor's voice in your head that sounds suspiciously like your second grade teacher who was sure you'd never amount to anything. Ha!

4. **Do write short bios for every major character, even though most of what you write in these bios will not make it into the narrative.** This is often the most important part of the process, because it unearths the motivation that lies beneath what your characters do, giving it meaning. It's what Fitzgerald meant when he so famously said, "Character is action"—meaning the things we do reveal who we are, especially because, as Gazzaniga reminds us, "Our actions tend to reflect our automatic intuitive thinking or beliefs."[13] Story is often about a protagonist coming to realize what's *really* causing him to do the things he does, at which point he either celebrates, because he's better than he thought, or begins making amends, because he's worse.

Developing an Outline: A Case Study

Now that we've talked the talk, let's walk the walk through the development of a set of interwoven thumbnail character bios that magically spawn a budding story outline.

THE PREMISE

Most writers begin with a premise; something along the lines of, "Hey, what would happen if . . . ?" A premise can be spurred by anything—something that happened in your life, something that leaps out when you're reading the newspaper, or even wishful thinking. To wit: you go to the movies. The hero is an actor who's getting a little long in the tooth (read: so old they probably aren't even his real teeth), yet his lead-ing lady has just finished cutting hers (read: is young enough to be his granddaughter). On the way home, you're bristling. How come when the man's much older, it's business as usual, but when the woman's older, it's *Harold and Maude*? (Forget *Cougar Town*.)

Of course, this wouldn't bother you so much if you weren't about to turn forty and, what's worse, you're mortifyingly infatuated with none other than Cal, the young actor who played the aging hero's son. Even the fantasy makes you blush. Then it hits you—you can't be more than thirteen, fourteen years older than Cal, whereas the age difference between the hero and the leading lady has to be at least double that. Is that fair? But since in real life there's not much you can do about it—after all, having things turn out exactly the way you want them to is pretty much wishful thinking—you're left with one surefire alterna-tive. Write a story.

So here is a fledgling premise: what happens when a woman about to turn forty meets the young actor she has a secret crush on and they fall madly in love?

Stop laughing. It could happen. The question is *how*. And no, we're not talking stalking, hypnosis, or a Vulcan mind meld. We're talking for keeps. *Voluntarily*. That means we have our work cut out for us.

ASKING OURSELVES, "WHY?"

On the surface, this story is about how a forty-year-old woman wins the heart of a twenty-six-year-old movie star. But what's it really about? What does winning the movie star's heart mean to her? What inner issue must she deal with in order to even try? To find out, we need to probe a little deeper. What's her love life like, anyway? Let's give her a boy-friend, one whose personality tells us something about her inner issue. How about a very nice but dull fiancé who's pressuring her to get mar-ried? And damned if she isn't considering it. Why? Because he's "safe." Does this mean she has a hard time taking risks? You bet. So what this story is *really* about is how a woman learns to overcome her fear of risk taking when she's forced to choose between a safe, comfortable future, and the possibility of an exhilarating one that comes with no guarantees.

Now we can take our premise—*Can a forty-year-old woman win the heart of a much younger man?*—and harness it to a theme: what happens when a person who's never taken a risk in her life throws caution—and a safe, comfortable life—to the wind? Which translates to: unless you take risks with the devil you don't know, chances are you'll spend the rest of your life shackled to the devil you do know. Now let's refine it a bit. What are we saying about human nature? How about: when you work up the courage to take a risk, good things happen, even if they're not quite the good things you expected. Great—now we have an idea of how the world is going to treat her.

So are we done with our character bio and outline? Nope. How do we know? Well, close your eyes. What do you see? Not much, which brings us to another handy one-step test.

HOW TO DIFFERENTIATE THE GENERAL FROM THE SPECIFIC IN A STORY

If you can't picture it, it's general. If you can see it, it's specific. As we'll explore in depth in chapter 6, you *must* be able to see it. The general, at best, conveys an objective idea that just sits there, idling in neutral; the specific personifies that idea, giving it a context that brings it to life. Big difference.

THE DETAILS

We still need to do more digging. For instance, this woman—let's call her Rae—what's her life like? Does she have kids? As a matter of fact, she does. A daughter. Is Rae divorced, then? Naw, don't want any ex-husbands lurking in the wings. Let's say Rae is widowed. Does she have a job? Nope. Her husband Tom left her enough to live on. Wait a minute, where's the goal in any of that? The conflict? It's inert. We're not looking for stats, we're looking for balls in play. So if her inner issue is that she isn't much of a risk taker, what in her past lets us know that? And hey, what triggered her warped worldview, anyway?

How about this: Rae wants to be a painter. Her mom was a painter; Rae learned at her feet. It thrilled her, the way everyone gushed over her mom's paintings—she didn't notice that no one ever actually offered to buy one. Until one day Rae overheard her mom's best friend talking with a neighbor about how horrendous everyone thought her mom's paintings were, but no one wanted to hurt her feelings by saying so. Rae was mortified for her mom, who she thought would be crushed if she knew. It wasn't a position she herself ever wanted to be in. So she's never shown a single one of her own paintings to anyone outside her family and friends. She *thinks* she has real talent. At least she hopes so. That's what keeps her going. Her fear is that she'll show her paintings to a pro and find out her main talent is the same as her mom's: self-delusion. Even so, she vows that soon she really will show them to the art dealer around the corner (aha, a goal!). But not today. That's been her plan for the past decade. Hey, it's worked so far.

Let's review. We know Rae's inner issue is her fear of risk. Thus, her closeted paintings now establish it as a "preexisting condition." And because it's specific, the reader figures that chances are it's something she'll try to overcome (meaning, it's something they can actively anticipate).

Next, let's turn our sights to Rae's daughter, whom we'll call Chloe. Why do we need her? No reason so far. The question, as with all subplots, is how does Chloe's existence impact the main storyline? Does it move it forward? Perhaps we should give Chloe a subplot that mirrors Rae's. We'll discuss subplots in depth in chapter 11; for now, suffice to say that mirroring subplots don't literally mirror the main storyline for the obvious reason—it would be redundant (hence boring). Instead, they reveal alternate ways in which the story question could be answered, usually for the protagonist's benefit—as either a cautionary tale or an incentive to change.

So how about this: Chloe is sixteen and plays the sax. She's good. So good she was just accepted to Juilliard, full scholarship. But because they live in, say, Charleston, South Carolina, it's a long way from home. This gives Rae several legitimate reasons why Chloe should stay home and finish high school, instead of skipping her senior year and moving to a strange city where she knows no one. Besides, despite what a great sax player Chloe is, there are no guarantees, and the life of a musician is so unpredictable. Chloe, of course, is dying to go. Will Rae let her?

Okay, we've set up a mirror. And something else—something you're always on the hunt for as you dig through your characters' backstories: current conflict. Especially conflict wired to a ticking clock. Like, say, that Chloe has a week to let Juilliard know whether she'll accept their offer. Good. Ball in play.

Now, what about Rae's dead husband, Tom? How does their relationship mirror or inform what will happen when she meets Cal? Well, here's a thought: since Cal is much younger than Rae, why not make Tom much older? Excellent choice. It means Rae knows a relationship with a large age gap can work—even though, of course, being the younger woman mitigated the risk that would have been involved on a more level playing field.

Which brings us to the force of opposition: what's standing in Rae's way (beside her inner issue)? Let's start with societal norms—the kind that spur the snickering assumption that a young man on the arm of a woman of a certain age means that money must be changing hands. Or worse, that she's a "cougar," conjuring the predatory image of heavy makeup, collagen-stung lips, and tummy tucks. This unspoken attitude permeates every element of the story, including Rae's psyche. Her heart beats with the question, *What will people say?* Look at what they said about her mom, and that was just over paintings.

Does that work as a force of opposition? Not yet. It's still too nebulous, too general. Sure, it will be reflected in the way certain characters react to Rae and Cal, but it remains conceptual. Close your eyes and you see nothing. We're looking for a more concrete obstacle, something we can picture. What Rae needs is a specific either/or, preferably one that will be affected by a possible relationship with Cal—which brings us to her boyfriend, the well-meaning but hapless Will, who has begun to push for marriage. Rae isn't sure why she hasn't said yes. He'd be a great stepfather for Chloe, he'd never stray, and he'd never tell her what to do. Which isn't to say there aren't a lot of traditional things he simply *expects* Rae to do. And why not? She's led a traditional life up to now. But what Will doesn't know is that the harder he pushes, the more she realizes there are other possibilities. They're just on the other side of a door she's never dared open. Risk. Then again, isn't security what everyone is really after? And Will isn't such a bad guy. So Rae promises to give him an answer by the end of the week.

Excellent. That's two balls in play.

And finally, what about Cal? What's his story? What's his goal? What's his internal issue? Story first: let's say Cal's been famous since he was fifteen. He's grown up in the spotlight. In two days he's due to begin filming the movie that will catapult him from megastar to icon—everyone says so. Trouble is, he's begun to suspect that being rich and famous isn't all it's cracked up to be, and he's feeling pretty darn sorry for himself. He's sick of being recognized wherever he goes. He wants to disappear for a few days, so he can decide what to do next. That's goal, internal issue, and ball three.

Okay. Now we know our major players. Are we ready to begin? Well, let's apply our eyes-wide-shut test. If you close your eyes, can you see anything yet? Nope. We're still backstage, in the dark. We have the Who and the Why. We need the Where and the How before the action can begin—aka the What Will Happen. The plot.

FIGURING OUT THE *WHAT*

Let's pry off another layer in search of a place where Rae and Cal might bump into each other. What if . . . there's a place each holds dear? What if it's the same place? Okay, that would work, but we have to be careful. It can't be the same place by coincidence—that is, because the plot needs it to happen. What we're looking for is a *story reason* that pulls them to the same place, at the same time. Fair enough.

What if Cal's family used to rent a vacation cottage every summer on a small rugged island off the Carolina coast? What if it was the last place he remembers being "himself," before fame struck? Okay, good.

Let's say Rae's had a crush on Cal since the first time she saw him on screen, when he really was jailbait, and long before he became as famous as he is now. That's why soon thereafter, when she read that Cal's family used to vacation on the island, on a lark she decided to see whether the cottage was still available as a summer rental. And guess what? It was. So for the past several years Rae, Chloe, and Will have summered there. Now we have something in both Rae *and* Cal's past that not only ties them to the same place, but ties them to it for the same reason.

Now that we have a logical place *where* Cal and Rae can believably be thrown together, what about the *how*? We don't want a lot of people gawking at them—not at first, anyway. In fact, best if these two can get to know each other alone. So let's sift through what we already know about them and see if we can come up with an answer.

What if . . . it's the end of the summer. Rae has a week to decide if she's going to marry Will and whether she'll allow Chloe to go to Juilliard. So she decides to stay on the island alone after everyone else has gone home, to make up her mind. She knows there's a bit of risk in this. The island will be deserted. And it's September, the middle of hurricane season. But after a lifetime of taking the safe route, she decides to take one little chance.

And didn't we give Cal a deadline, too? He's supposed to report to the set for that blockbuster he's about to film. But like Rae, he is having second thoughts about his future. Knowing his life will be forever changed if he appears in the movie, he needs to take a time out. He needs to be alone to figure out what to do next. And what better location than the last place he remembers being happy? The island. After all, it'll be deserted; how hard can it be to break into the cottage where he stayed as a boy?

Notice that both our main characters have a clock that just started ticking. That means we've found our beginning. Each one is standing on the shore of "before," staring into the distance, trying to make out the shape of "after." The story will chart the path in between.

Now we have Why, Where, How, When, and Who. Close your eyes and you can begin to see it actually unfold. Is it the kind of perfectly formatted hierarchical outline that would've received a gold star back in elementary school? Probably not. Is it enough for you to start writing? Quite possibly. Our story is now securely anchored in the "before," and what happens will be meted out by ticking clocks tied to specific upcoming events, which will force our protagonists to confront the long-standing fears and desires that up to now they've swept under the rug. There will be a mounting sense of urgency, and readers will indeed be able to anticipate what happens next.

Do we know the answer to our original premise: what happens when a woman about to turn forty meets the young actor she has a secret crush on, and they fall madly in love? Nope. We know something even more important. Turns out that's not what our story is about. It's really about whether Rae can overcome her fear and risk showing her paintings, knowing that, regardless of the reaction, she'll be okay. It's about facing who you are and taking the consequences, not to mention the perks, one of which just might be finding your true love. Just saying.

Have we set the stage to find out? Yep. So you see, outlining doesn't have to take the spontaneity out of writing. You don't need to know *exactly* how the story is going to end, but you do need to know what the protagonist will have to learn along the way—that is, what her "aha!" moment will be. And even if you do have a precise scene-by-scene outline? As we discussed in chapter 2, there's no law that you have to stick to it. Sometimes the excitement of writing is discovering those places where the story suddenly careens into new territory on its own—and you realize its new direction makes even more sense than the one in which it was headed. Of course, in this as in most things in life, luck tends to favor the prepared.

And the best preparation for writing any story is to know with clarity what your protagonists' worldview is, and more to the point, where and why it's off base. Thus you have a clear view of the world as your protagonist sees it and insight into how she therefore interprets, and reacts to, everything that happens to her. It's what allows you to construct a plot that forces her to reevaluate what she was so damn sure was true when the story began. That is what your story is really about, and what readers stay up long past their bedtime to find out.

CHAPTER 5: CHECKPOINT ✓

Do you know why your story begins when it does? What clock has started ticking? What is forcing your protagonist to take action, whether she wants to or not?

Have you uncovered the roots of your protagonist's specific fears and desires? Do you know what her inner issue is? Can you trace it all back to specific events in her past? Do you know how her inner issue then thwarted her desire right up to the moment the story begins?

Have you made your characters reveal their deepest, darkest secrets to you? I don't want to go all Big Brother on you, but if you let your characters hold back, we'll know. Trust me.

When writing character bios, are you being specific enough? When you close your eyes, can you envision what happens, or is it still conceptual? If you can't see it, there will be no yardstick to measure your protagonist's progress. You can't have an after without a before.

Do you know where the story is going? This isn't to say you need to know how it ends when you write word one (although it's not a bad idea), but unless you have *some* clue where it's headed, how can you be sure you've sown the seeds of the future there on page one?

6

THE STORY IS
IN THE SPECIFICS

COGNITIVE SECRET:
We don't think in the abstract; we think in specific images.

STORY SECRET:
*Anything conceptual, abstract, or general must be
made tangible in the protagonist's specific struggle.*

Advice to young writers who want to get
ahead without any annoying delays: don't
write about Man, write about a man.

—E. B. WHITE

WAIT, I HEAR YOU SAYING. Some people think in the abstract. Scientists, mathematicians, braniacs like Albert Einstein, for instance. He didn't arrive at things like $E = mc^2$ by channeling Jane Austen. No, he came up with it after remembering how, as a child, he'd imagined riding through space on a beam of light. And relativity theory? By imagining what it would be like to plummet down an elevator shaft, then take a coin out of his pocket and try to drop it—without, I'm assuming, passing out or throwing up first. Here's how Einstein explained his own mental process: "My particular ability does not lie in mathematical calculation, but rather in visualizing effects, possibilities, and consequences."[1]

Sounds exactly like a story to me. And the key word here is *visualizing*. If we can't see it, we can't feel it. "Images drive the emotions as well as the intellect," says Steven Pinker, who goes on to call images "thumpingly concrete."[2]

Abstract concepts, generalities, and conceptual notions have a hard time engaging us. Because we can't see them, feel them, or otherwise experience them, we have to focus on them really, really hard, *consciously*—and even then our brain is not happy about it. We tend to find abstract concepts thumpingly boring. Michael Gazzaniga puts it this way: "Although attention may be present, it may not be enough for a stimulus to make it to consciousness. You are reading an article about string theory, your eyes are focused, you are mouthing the words to yourself, and none of it is making it to your conscious brain, and maybe it never will."[3]

Story, on the other hand, takes mind-numbing generalities and makes them specific so we can try them on for size. Remember, we're hardwired to instantly evaluate everything in life on the basis of *is it safe or not?* Thus the whole point of a story is to translate the general into a specific, so we can see what it really means, just in case we ever come face to face with it in a dark alley.

And the only way we can see it, is if we can, well, *see* it. As Antonio Damasio says, "The entire fabric of a conscious mind is created from the same cloth—*images.*"[4] Neuroscientist V. S. Ramachandran agrees: "Humans excel at visual imagery. Our brains evolved this ability to create an internal mental picture or model of the world in which we can rehearse forthcoming actions, without the risks or penalties of doing them in the real world."[5] What this all boils down to is, as I'm inordinately fond of saying, the story is in the specifics.

Yet writers often tell entire stories in general, as if concepts alone are captivating or, worse, because they've fallen prey to the misconceived notion that it's the reader's job to fill in the specifics. That's why in this chapter we'll explore the difference between the specific and the general; why the specific often turns up missing; where writers often inadvertently drop the ball; and why giving *too many* details is just as bad as not giving enough. Finally, we'll tackle the myth that sensory details inherently bring a story to life.

The General Versus the Specific

In October 2006, nearly six thousand people worldwide perished in hurricane-induced floods.

Quick, what do you feel after reading that sentence?

My guess is, you feel a little confused by the question.

Now imagine a wall of water coming straight toward a small boy, who clings desperately to his frantic mother. Trying to soothe him, she whispers, "Don't worry baby, I'm here, I won't let you go." She feels him relax in the moment of deafening calm just before the water rips him from her arms. The sound of his cry above the cacophony of destruction—trees ripped from the ground, houses smashed to splinters—will haunt her for the rest of her life. That, and his look of utter surprise as he was swept away. *I trusted you,* it seemed to say, *and you let me go.*

Now how do you feel? This time, the question is clear. Watching the flood claim that one little boy is far more gut-wrenching than hearing about six thousand anonymous people perishing in various floods, isn't it? I'm not suggesting your heart doesn't go out to all the flood victims and their families. But chances are, when you read that opening sentence, you didn't feel much of anything.

Don't worry. This isn't a psychological test to reveal your deep-seated pathological tendencies; rather, it highlights how we humans process information. As counterintuitive as it may seem, even the most massive, horrendous event, when presented in general, doesn't have much direct emotional impact, so it's easy to sail right by it almost as if it wasn't there. Why? Because we'd have to stop and *think* about it in order to "manually" do what a story would have done: make it specific enough to have an emotional impact. And why would you do that? As Damasio says, "Smart brains are also extremely lazy. Anytime they can do less instead of more, they will, a minimalist philosophy they follow religiously."[6] Since your brain's probably much more interested in thinking about something that matters, like why your spouse is late again tonight, it's probably not going to work at envisioning—wait, what was that again? A terrible flood somewhere years ago? Especially because hey, there's nothing you can do about it, and besides, it would just make you feel bad, and god knows you have enough on your plate with your knucklehead spouse, who your mother warned you about, but did you listen? *Huh?* Flood? You talking to me?

The point is, if I ask you to think about something, you can decide not to. But if I make you feel something? Now I have your attention. Feeling is a reaction; our feelings let us know what matters to us, and our thoughts have no choice but to follow.[7] Facts that don't affect us—either directly or because we can't imagine how the facts affect someone else—don't matter to us. And that explains why one personalized story has infinitely more impact than an impersonal generalization, even though the scope of the generalization is a thousand times greater. In fact, it is only via a specific personalization that the point of a generalization is shot home. Otherwise, as Scarlett said, we can think about it tomorrow—which, given how much brain energy it takes to think about something that hasn't grabbed us emotionally, usually translates to a week from never.

Feel first. Think second. That's the magic of story. Story takes a general situation, idea, or premise and personifies it via the very specific. Story takes the horror of a huge, monstrous event—the Holocaust—and illustrates its effect through a single personal dilemma—*Sophie's Choice*. Thus the massive, unwieldy, unbearable vastness of its otherwise incomprehensible inhumanity is filtered through its effect on one person, a mother who must decide which of her two beloved children to spare. And because we are in Sophie's skin, we feel the ineffable magnitude of all of it: the Holocaust, the unspeakable cruelty, her ultimate decision. We are not just being *told* about its effect; we are experiencing it.

The Specifics About Specifics

But to unearth the generalities that can undermine a story, we need to know what they look like. The answer is simple: a generic doesn't look like anything at all, which is the point. A generic is a general idea, emotion, reaction, or event that does not refer to anything specific. For instance, telling us "Trevor had a great time," without telling us what Trevor actually did, or what he considers to be a great time, is generic. Telling us, "Gertrude always wanted to start her own business," without telling us what that business is, why it's interesting to her, and why she hasn't, in fact, started it already, is generic. Generic concepts are crafty devils. They leap in front of your story and pull the blinds down, shutting the reader out. Here's a specific example of how maddening generics can be when they sneak into a story and take hold:

JAKE

Kate, we've been working together a long time.

KATE

Eons.

JAKE

And I've come to expect a certain, oh, how shall I say it? *je ne sais quoi* in your work.

KATE

Thank you Jake. I think.

JAKE

Unfortunately, your work on this project has been subpar.

KATE

But I put everything I've got into it.

JAKE

I'm not questioning how hard you've worked. I question your technique and lack of progress. Have you forgotten this is the firm's most prestigious project? Everything's riding on it. I'll give you a few days, but if you don't produce, I'll have to transfer you back to your old job.

KATE

I can't believe you'd even consider that, given what happened last April.

JAKE

My point exactly! Now, get back to work before I regret my decision.

Clearly, as far as the writer is concerned, these two characters are in the midst of an intense, conflict-driven turning point. It's easy to picture the writer as her fingers fly over the keys, sure she's giving voice to Kate's growing anxiety and Jake's measured frustration. And sure enough, we feel anxiety and frustration, too, because we have absolutely no idea whatsoever what Kate and Jake are actually talking about.

CASE STUDY: WALLY AND JANE

Let's deconstruct a similarly vague sentence to give us an even better idea of what, specifically, vague looks like:

> Jane knew Wally had a reputation for doing horrid things, so when he commented on her appearance in front of everyone, she refrained from smacking him.

On the surface this sounds like a perfectly reasonable sentence, that is, as long as the *next* sentence goes on to answer the questions this one raises. Unfortunately, what tends to follow is usually another sentence just as full of vague generalities. With that in mind, let's take a good hard look at what that sentence *doesn't* tell us:

Not only don't we know what kind of horrid things Wally has done, we don't know what Jane would *see* as horrid. For instance, perhaps Wally sets stray cats on fire. That would be pretty horrid. And it would tell us something about Wally. Or maybe Wally hangs out with poor kids from across the tracks, which Jane and her stuck-up posse think is absolutely unforgivably horrid. That would tell us something about both Wally *and* Jane.

And given that this happened in front of "everyone," how did they react? Well, that depends on who they are. Are they steel workers? High school students? Strangers in the subway? And even if we knew *exactly* who they are, we couldn't so much as guess how they'd react to Wally's comment, because we have no idea what the comment was.

But before we get to what Wally said, there's that word, "comment." Wally "commented" on her appearance. Comment as in diss? Comment as in a come-on? We don't know. All we know is that Jane had a strong reaction to it. Did he ask if she put on weight? Did he tell her that if she doesn't want him staring at her breasts, she shouldn't wear a skin-tight, low-cut baby T-shirt that says "Juicy" in rhinestones across the front? Or maybe her desire to smack him stems from the fact that he talked to her at all, given that she's the senior homecoming queen and he's a geeky grease monkey. We don't know what the truth is, so even if we try to make an educated guess, *we have no way of knowing whether we're right*. So, regardless of what we come up with, it'll feel like picking a number out of a hat—and be just about as satisfying.

The same confusion comes up around the smack. Did Jane refrain from smacking Wally hard across the chops? Or would it have been a playful pat on the butt? Or is *smack* slang for *kiss*, 'cause what he actually said was, "Babe, you look gorgeous," which was music to her ears because she's had a crush on him from the moment she heard he sets cats on fire, since as it turns out, that's her secret hobby, too? The possibilities, as Buzz Lightyear would say, go to infinity and beyond. Which puts chances at next to nil that the reader would come up with the right answer:

> Jane knew Wally liked to eat worms so he could gross everyone out by barfing them up during show and tell, so when he called her a sissy in front of the whole kindergarten class, she decided not to punch him in the stomach and give him the pleasure.

The problem with generalities is that because they're utterly ambiguous, they don't have legs. Because they don't tell us specifically what is happening now, we can't anticipate, specifically, what might happen next. So much for the delicious dopamine rush of curiosity that keeps us reading.

The point is, generalities are not capable of producing specific *consequences,* and so the story has nowhere to go. Instead, more vague things happen, compounding the confusion, until the reader realizes that she has far more questions than the story will ever answer and heads to the refrigerator for a snack.

Why Would a Writer Be Vague?

Writers are rarely aware they're being vague, although as we'll see in the following list, sometimes they actually do it on purpose. They tend to entrust their story to generics for three main reasons:

1. The writer knows the story so well, she doesn't recognize when a concept that's very clear to *her* will come across as utterly opaque to the reader. So when she writes, "Renee looked at Osgood in his tight jeans, tousled hair, and ratty Converse high tops, and smiled knowingly," she has absolutely no idea that it leaves us thinking, *What do you mean, knowingly*? What's behind that smile? Her knowledge that Osgood's really a pretentious poser rather than the guileless hipster he pretends to be? That he's her dream guy and tonight's the night she's going to tell him? That she's pregnant with Axel's baby, but Osgood will never be the wiser? It doesn't even occur to the writer to tell us; because *she* knows exactly what "smiled knowingly" refers to, she assumes we do, too.

2. The writer *doesn't* know the story well enough, so when Renee tosses her head and gives Osgood that knowing grin, it's because the plot needs her to. If asked about it, the writer might look at you quizzically and say, "Wait, you mean she needs more of a reason than that?"

3. The writer knows her story very well and is quite aware she hasn't told the reader what's behind Renee's knowing smile, because she's afraid if she does, she'll "give too much away."

This oft-misguided fear is something we'll be talking about in depth in chapter 7 when we discuss "reveals"—and so, uh, I don't want to give too much away here.

Whether it stems from the writer knowing too much or too little, or actually doing it on purpose, being vague is never a good idea. So to help you zero in on wherever vagueness may have crept into your story, here's a rundown of where the usual suspects tend to lurk.

Six Places Where the "Specific" Often Goes Missing

1. **The specific *reason* a character does something.** Like most things, it can start off so promisingly: "Holly ducked into the alley, glad to have avoided Sam for the millionth time." Sounds great, right? Trouble is, unless we know at that moment in the story *why* Holly has been avoiding Sam, it will fall flat. It could be because he's been stalking her since 1967 or she's secretly in love with him and doesn't want him to see her on yet another bad hair day or she owes him money. Who knows? Each of these specific possibilities suggests a very different scenario, any one of which would help us make sense of what's happening in the moment and allow us to anticipate what might happen next. Without a specific, we have no clue.

2. **The specific *thing* a metaphor is meant to illuminate.** Here's an interesting fact to add to what we already know: not only do we think in story and in images, but as cognitive linguist George Lakoff points out, although we may not always know it, we also think in metaphor.[8] Metaphor is how the mind "couches the abstract concepts in concrete terms."[9] Believe it or not, we utter about six metaphors a minute. Prices *soared*. My heart *sank*.

Time *ran out*. Metaphor is so ubiquitous we rarely notice it's there.[10] Ah, but literary metaphor is something else again—it's intended to convey new insight. Literary metaphor isn't hidden; its point is to be recognized as such. To quote Aristotle's perfect definition: a "metaphor consists in giving the thing a name that belongs to something else."[11] The trouble is, sometimes the writer gets so carried away with crafting a beautifully written, evocative metaphor, he forgets to tell us what, exactly, the "thing" it's being compared to actually is. Here's an example:

> Something deep inside Sam was about to tear; he felt it pulling apart at the seams. He pictured it like a clumsy teen's well-used softball, the stitching now a grimy gray. Once that stitching pops, though, it will become something else, as the cover peels off revealing something ugly and strange, something you would never suspect had always been at the heart of that once gleaming, achingly hopeful softball.

It's evocatively written, but because we have *no idea* what the "something ugly and strange" actually corresponds to in the story, or what point the author is trying to make other than that something vague and unspecified inside Sam is going to rip like a softball, it is also uninvolving. Metaphors have resonance only when we know, specifically, what they're supposed to illuminate. Otherwise, although it definitely sounds like it's meant to tell us something really important, we're left thinking, *I know this has great significance, but I have no idea of what.* Nor should we have to spend even a nanosecond decoding a metaphor. It should be "gettable" when reading at a clip, and its meaning instantly grasped. What's more, metaphors need to give us new information and fresh insight rather than simply restating something we already know, no matter how poetically.

3. **The specific *memory* that a situation invokes in the protagonist.** Here's another great start:

> The minute Sam threw the stinky old softball at Holly, he knew it was a mistake. If only he'd learned his lesson during that unforgettable eleventh inning at Lake Winnatonka Camp for the Clumsy during the summer of 1967—but sadly, no, he hadn't.

We're left thinking, *Wait, what lesson? Why was it unforgettable?* Because without the specifics—what *actually happened* back in 1967—we have no idea what Sam should have learned from it, how it applies to what's happening now, or what it's meant to tell us about the dynamic between Sam and Holly. Because the reader has no point of reference, the best she can do is make up something. This is even more maddening than it sounds, because she'd then have no way whatsoever of knowing whether she was right. And worse, since the chance of her actually envisioning the specifics the writer left blank are about as likely as either of us winning the lottery, she's now imagining a decidedly different story than the one the writer actually wrote.

4. **The specific *reaction* a character has to a significant event.** Let's follow Holly and Sam a bit longer:

> Sam was terrified that if Holly spotted him following her again with the softball in his pocket, she'd not only nix their spaghetti dinner rendezvous that night but would finally take out that restraining order. He was so worried about it that he didn't notice she'd stopped to tie her shoe, and he tripped over her. Now she knew he'd been tailing her, there was no getting around it.
>
> The next day Sam went to work, hoping his boss was in a good mood, because he wanted to ask him about that promotion. . . .

And we're left thinking, *Hey, wait a minute, wasn't Sam worried about what would happen when Holly found out he was following her? What conclusion does he draw? What's the result? The consequence? How does he feel? Say something, anything!* What's worse, because we knew Sam was extremely concerned about it, the fact that he's not reacting an iota makes us wonder if he's made of flesh and blood after all. Hey, maybe aliens really are among us.

I know this example seems extreme, but it's astoundingly common. Why? My guess is that since the writer clearly told us how much Holly meant to Sam, she figured we'd know exactly how Sam felt, so why should she have to waste time spelling it out? But although we can indeed imagine how Sam feels in general—say it with me—the story is in the specifics.

The point is, characters need to react to everything that happens for a specific reason we can grasp in the moment. Of course, there may be a deeper reason as well that we won't fully understand until later. In fact, the "real reason" for a reaction may be the opposite of what it looks like now. But what there can't be, if you want your readers to stay with you, is *no* reaction. This is especially true when we've been led to believe that a character will be hugely affected by something that then doesn't cause him to bat an eyelash. It's one more reason to always keep in mind that the story isn't in what happens; it's in how your characters react to it.

5. **The specific *possibilities* that run through the protagonist's mind as she struggles to make sense of what's happening.** This is ripped straight from the pages of our generic story:

> Holly realized Sam had been stalking her all these years. *Why on earth would he do that, and what's up with that softball?* she wondered, racking her brain for an answer, but not coming up with anything that would explain it.

This time we're left thinking, *Wait; can you at least tell us what the options were? What went through Holly's brain as she racked it?*

For a look at just how much information you can convey in a mental passage before your protagonist delivers an actual response, here's a revealing snippet from Eleanor Brown's *The Weird Sisters:*

> She remembered one of her boyfriends asking, offhand-edly, how many books she read in a year. "A few hundred," she said.
>
> "How do you find the *time*?" he asked, gobsmacked.
>
> She narrowed her eyes and considered the array of potential answers in front of her. Because I don't spend hours flipping through cable complaining there's nothing on? Because my entire Sunday is not eaten up with pre-game, in-game, and post-game talking heads? Because I do not spend every night drinking overpriced beer and engaging in dick-swinging contests with other financi-rati? Because when I am waiting in line, at the gym, on the train, eating lunch, I am not complaining about the wait/ staring into space/admiring myself in available reflective surfaces? I am *reading*!
>
> "I don't know," she said, shrugging.[12]

Need I say more?

6. **The specific *rationale* behind a character's change of heart.** Meanwhile, back at our generic story:

> Once she realized Sam was following her, Holly vowed that if there was one thing she'd *never* do, it was have spaghetti with him. But when he texted her that the water was boiling and she had eight minutes to get to his house before the pasta got mushy, after a raging internal

debate, Holly texted back, "Yes, I love it al dente, I'll be there in five."

By now you know the million-dollar question: *why* did Holly change her mind? And the answer can't be "just because." We want to be privy to her raging internal debate, and what it is that finally tips the scales.

Specifics Are Good, but Less Is Often More

Before we get carried away and load up our stories with specifics as if they're plates at an all-you-can-eat buffet, it pays to keep Mary Poppins' sage advice in mind: enough is as good as a feast. Too many specifics can overwhelm the reader. Our brain can hold only about seven facts at a time. If we're given too many details too quickly, we begin to shut down. For instance, can you make it to the bottom of the following paragraph?

Jane glanced into the yellow room, her gaze quickly taking in the massive four-poster bed with the fluffy blue-and-green paisley quilt, the craftsman rocker, the matching oak end table, laden with books, dust, and a huge brass lamp with flickering flame-shaped bulbs, ominously teetering next to sixteen unpacked fraying brown boxes, the one nearest to the door full of old clothes from the sixties—leather mini-skirts, muslin halter tops, skin-tight knee-high crinkly white patent leather boots, yellow Mary Janes, bellbottom jeans, and a floppy purple suede cowboy hat—the other fifteen boxes containing everything that Matilda had collected during her very long life, and if Matilda was anything, she was a packrat, so there was. . . .

Quick, what color was the room? If you're thinking, *What room?* I don't blame you. I'm guessing your eyes glazed over about three lines down. Because although the writer may have known why each detail was important, the reader doesn't have a clue. And we can't even pause to try to figure it out, because the details keep on coming. So by the end of the paragraph, we have lost track of not only the details, but the story itself.

Think of each detail as an egg. The writer keeps tossing them at us, one after another, seemingly unaware of the growing number of precariously balanced eggs we're being asked to hold. So somewhere around the middle of the description—say, the huge brass lamp—it's one egg too many. The trouble is, we don't just drop that particular egg; *all* the eggs go crashing to the ground. The more details the writer gives us, the fewer we'll remember, proving, once again, that as with most things in life, less is more. Take it from iconic singer Tony Bennett who, when asked what he can put into a song in his eighties that he couldn't when he was younger, answered without missing a beat, "The business of knowing what to leave out."[13] Why wait until you're eighty to master that one?

But, popular wisdom goes, there's one type of detail you can never have too many of: sensory details. Writers are advised to infuse their stories with abundant, sun-drenched, crunchy, tactile, savory sensory details, the better to draw readers into the story.

Really?

MYTH: Sensory Details Bring a Story to Life

REALITY: Unless They Convey Necessary Information, Sensory Details Clog a Story's Arteries

Like everything else in a story, details and specifics need a *story reason* to be there. This is especially true of sensory details. I remember reading the first page of a manuscript that waxed eloquent about how the warmth of the sun felt on the back of the protagonist's hands as she drove down a quiet early morning lane, the way the taste of the sumptuous strawberry she'd eaten for breakfast lingered on her tongue, how the coolness of the steering wheel beneath her palms made her shiver with delight . . . and that's about all I remember because by then all I could think about was how refreshing a nice little catnap would be.

Just because the sun beat down on the protagonist's skin, we don't need to know it. Just because she could still taste the strawberry despite the fact she'd brushed her teeth, flossed, and gargled six times, we don't need to know it. Just because the steering wheel felt cool to the touch, well, you know the drill. We need to know these things *only* if they supply a necessary piece of information. For instance, let's say the protagonist—we'll call her Lucy—thrills to the cool, pure sweetness of a rich vanilla malted. Who cares, right? Unless, with one last sip, Lucy passes out because she's hypoglycemic—bingo, a consequence. It would be even better if it also gave us insight into Lucy, so perhaps by having her suck down the malted, the writer is telling us Lucy's a hedonist who puts momentary pleasure over her long-term health. Or, just as effective, maybe Lucy's love of vanilla is making the metaphorical point that while all the women in the steno pool simply adore chocolate, Lucy's allegiance to vanilla implies that she's not one of the pack and shudders at the marginalizing assumption that all women love chocolate. *Hell,* the reader might then think, *I bet Lucy doesn't have a closet full of impulse-buy shoes either or spend all her spare time getting facials and catching up on key celebrity gossip.*

As Chip and Dan Heath point out in *Made to Stick,* while vivid details can boost a story's credibility, they must be meaningful—that is, they need to symbolize and support the story's core idea.[14] Remember those 11,000,000 bits of information our five senses are lobbing at us every second? *They* are sensory details. Yet our brain knows that we need to be shielded from at least 10,999,960 of them. The only details it lets through are the ones with the potential to affect us. The same is true of your story. Your job is to filter out the details that don't matter a whit so you can have plenty of space left for the ones that do.

There are three main reasons for any sensory detail to be in a story:

1. It's part of a cause-and-effect trajectory that relates to the plot—Lucy drinks the shake, she passes out.

2. It gives us insight into the character—Lucy's an unapologetic hedonist headed for trouble.

3. It's a metaphor—Lucy's flavor choice represents how she sees the world.

In addition, the reader must be *aware* of the story reason for each detail's presence. Plot-wise, that's a no-brainer: while in the midst of savoring the malt, Lucy slips from consciousness and flops to the floor. The connection would be hard to miss. As for its telling us she's a hedonist, first we'd need to know she's hypoglycemic and is fully aware of the danger lurking in that innocent-looking malted. This is the sort of setup writers often overlook in a first draft, but which can easily be inserted in the second.

The third option—that it tells us something metaphorically—is the trickiest to convey. That's because it doesn't rely on something concrete—either physical action or our awareness of a specific fact—to make sense. Rather, it depends on the reader's ability to grasp its subtext. That, in turn, depends on the writer's ability to have laid the groundwork so we can intuit that Lucy's chosen flavor of malt reveals the fact that she dances to the beat of her own drummer. Thus the reader would already need to know that the rest of the steno pool sees loving chocolate as something that defines their identity, which corresponds to a cultural conformity that Lucy finds stifling. *Lucy looked around the diner at all the women slowly sipping chocolate malts as if it was some kind of secret handshake, entrée into a club that she had no desire to join.* And so drinking the vanilla malted turns out to be a very courageous act indeed because of what it reveals about her character—it says Lucy is a woman with the courage of her convictions. A piece of information that, from then on, will color the reader's take on everything Lucy does and every situation she finds herself in.

Location, Location, Location

Writers hoping for an exception to the needs-a-story-reason-to-exist rule often point to scenery. I mean, we have to know where the story takes place, right? The layout of the bedroom, the sagging porch floorboards, the weeping willow in the yard, the soaring mountain ranges—and who doesn't love a beautiful sunset? As Elmore Leonard so shrewdly advised, "Try to leave out the part that readers tend to skip."[15] And a large part of what readers skim, if not skip entirely, is scenery. Setting. Weather. Why? Because stories are about people, the things that happen to them, and how they react to it. And while setting is where those things take place, so of course it's vitally important, merely describing the scenery, the town, the weather—regardless of how well written or how interesting it might be in and of itself—stops a story dead in its tracks.

This isn't to say that you can't mention the gothic architecture, that it was a dark and stormy night, or that the town dates back to 1793. But when you do, it helps to keep George S. Kaufman's old Broadway saw in mind: "You can't hum the scenery." We need a story reason to care how ominous the clouds are, how vibrant the city, how quaint the white picket fence. Often, description of the scenery sets the tone. As Steven Pinker says, "Mood depends on surroundings: think of being in a bus terminal waiting room or a lakeside cottage."[16] So if you go to great pains to describe the scenery—be it a room, a setting, an elaborate meal, or what your protagonist is wearing—you'd better actually be communicating something else. The description of a room should reveal something about the person who lives in it or hint at the whereabouts of the missing diamond or tell us something crucial about the zeitgeist of the community in which the story unfolds—or better yet, all three.

For example, let's turn to a master, the unparalleled Gabriel Garcia Marquez, and crib something from *Love in the Time of Cholera*. The following paragraph is a quintessential example of how the mere description of a room can be harnessed to provide insight into a character. In this passage, Dr. Juvenal Urbino surveys the parlor of his good friend and chess partner, photographer Jeremiah de Saint-Armour, who has just committed suicide.

> In the parlor was a huge camera on wheels like the ones used in public parks, and the backdrop of a marine twilight, painted with homemade paints, and the walls papered with pictures of children at memorable moments: the first Communion, the bunny costume, the happy birthday. Year after year, during contemplative pauses on afternoons of chess, Dr. Urbino has seen the gradual covering over of these walls, and he had often thought with a shudder of sorrow that in the gallery of casual portraits lay the germ of the future city, governed and corrupted by these unknown children, where not even the ashes of his glory would remain.[17]

The snippet not only provides backstory and insight into how Dr. Urbino views the world, it also deftly sums up an aspect of the universal human condition that all of us struggle with—that someday the world will go on without us, perhaps as if we had never been. Which, ahem, is one of the reasons we write stories. It's better than spray painting "Kilroy was here" on a rock.

So if you want to write a novel that inspires people you've never even met to call their friends and say, "You *gotta* read this book," you need to root through your story and make sure you've translated anything brain-numbingly vague, abstract, or generic into something that's surprisingly specific, deliciously tangible, and grippingly visceral.

CHAPTER 6: CHECKPOINT

Have you translated every "generic" into a "specific"? This is another way of saying, "Do your job." After all, you don't want your reader filling in the blanks in ways you never intended.

Have the specifics gone missing in any of the usual places? Are there places where the reason, rationale, reaction, memories, or possibilities that underlie your protagonist's actions are invisible to the reader?

Can your reader see what, specifically, your metaphors correlate to in the "real world," grasp their meaning, and picture them, when reading at a clip? The last thing you want is for your reader to have to reread it three or four times—first to be able to picture it and then to figure out what the heck it means.

Do all the "sensory details"—that is, what something tastes, feels, or looks like—have an actual story reason to be there, beyond "just because"? You want to be sure each sensory detail is strategically placed to give us insight into your characters, your story, and perhaps even your theme. And remember, scenery without subtext is a travelogue.

7

COURTING CONFLICT, THE AGENT OF CHANGE

COGNITIVE SECRET:

The brain is wired to stubbornly resist change, even good change.

STORY SECRET:

Story is about change, which results only from unavoidable conflict.

All changes, even the most longed for, have their melancholy, for what we leave behind us is a part of ourselves; we must die to one life before we can enter into another.

—ANATOLE FRANCE

THE BRAIN DOESN'T LIKE CHANGE. Would you, if you'd spent millions of years evolving with the sole goal of maintaining a constant, stable equilibrium? And the brain didn't slack off after mastering mere physical survival, no sir; it turned its sights to making sure we had a nice comfortable sense of well-being to go along with it. Only then did the brain settle in for the long haul—unseen yet vigilant—ready to pounce on any possible imbalance, often before it hits our conscious radar.[1] *That* explains why the thought of switching barbers, taking a new route to work, or buying a different brand of toothpaste can feel disconcerting enough to keep us loyal to our old brand. After all, our teeth haven't fallen out yet, so it must be working. Why rock the boat?

And as neuroscience writer Jonah Lehrer points out in *How We Decide*, "Confidence is comforting. The lure of certainty is built into the brain at a very basic level."[2] In fact, it's a big part of our sense of well-being. That is why, when questions arise that challenge our beliefs about, well, anything, we tend get a little cranky. Or as social psychologist Timothy D. Wilson says, "People are masterful spin doctors, rationalizers, and justifiers of threatening information and go to great lengths to maintain a sense of well-being."[3]

We don't like change, and we don't like conflict, either. So most of the time we do our best to avoid both. This isn't easy, since the only

125

real constant *is* change, and change is driven by conflict. *This or that? Me or you? Chocolate or vanilla?*

Sounds sort of bleak, doesn't it? *But wait*, as they say on late night infomercials, *there's more!* As anyone who's ever fallen under the spell of a sparkly bauble, charming stranger, or cockeyed dream knows, there's another side to this coin. The lure of the new, the novel—of that bright shiny thing hovering just out of range—is equally hardwired.

We evolved as risk takers, too. We had to. Without a sense of adventure, we wouldn't have gone off in search of the wild prey that fed our growing brain, dared to scale the mountain range that led to the life-sustaining verdant valley below, or had the nerve to approach that charming stranger who then made life worth living.[4]

And there's the paradox: we survived because we're risk takers, but our goal is to stay safe by not changing an iota unless we absolutely have to. Talk about conflict! And that brings us right back to *story*. Story's job is to tackle exactly how we handle that conflict, which boils down to this: the battle between fear and desire.

Thus it's no wonder that from time immemorial conflict has been called the lifeblood of story. It's something everyone tends to agree on, whether cranking out a mass market potboiler about killer spiders or penning an exquisitely rendered literary novel that turns on whether or not the protagonist sighs when the postal carrier at long last slips the much-anticipated ivory envelope through the tarnished brass mail slot. As a result, creating conflict seems delightfully clear, completely up front, embarrassingly obvious.

Well, I'm here to tell you not to believe that for a minute. Instead, my friends, consider conflict the original passive-aggressive devil. So in this chapter, in the hope of wrestling that devil to the mat, we'll explore how to harness impending conflict to mounting suspense from the very first sentence; where the specific avenues of conflict and suspense are often found; and why holding back crucial information for a big reveal later ironically tends to ensure that readers will never get there.

Understanding Our Conflict with Conflict

When it comes to conflict, your reader—like the pasty-faced kid in *The Sixth Sense*—must be able to see things that aren't there. In order for readers to sense that "all is not as it seems," conflict must be palpable long before it rises to the surface. It's the *potential* for conflict that gives urgency to everything that happens, underscoring even the most benign events with portent. Indeed, it ripples through the story in the guise of mounting tension, engendering in the reader that delicious dopamine-driven sensation we're addicted to when it comes to a good tale: suspense, the desire to find out what's really going on.

But when it comes to portraying conflict on the page, how we're wired for real life tends to muck things up. "Since we are social creatures, a need to belong is as basic to our survival as our need for food and oxygen," says neuropsychiatrist Richard Restak.[5] It started a couple of hundred thousand years ago when it first dawned on us that, survival-wise, two heads are better than one, and a whole society, better yet! Thus a new human goal was born, one still championed by kindergarten teachers the world over: working well with others. This gave rise to a whole host of emotions—some pleasant and some decidedly not—to encourage us to get along. And for anyone with lingering doubts about the unequaled power of emotion, a recent study using magnetic resonance imaging revealed that intense social rejection activates the same areas in the brain that physical pain does.[6] Our brain is making a point. Conflict hurts.

That's probably why we try to defuse conflict as quickly as possible. We are made to understand at a very early age that we bring conflict into relationships at our own peril, and we are rewarded—by both society and the chemicals in our brain—for finding ways to nip it in the bud before it escalates. As the old song goes, the idea is to focus on the positive rather than the negative and—whatever you do—"don't mess with Mr. In-Between."[7] The thing is, *every* story tells the tale of Mr. In-Between. As in, between a rock and a hard place. Yet it's so easy to fall prey to our unconscious urge to steer clear of rocks and hard places and, in accordance with the golden rule, to never put anyone *else* in that position, either—including, unfortunately, our protagonist.

I'll never forget working with an author who had written an eight-hundred-page manuscript, a novel about a guy named Bruno, chronicling his rise from poverty to ill-gotten riches as a ruthless Mafia don. Or make that, he *probably* would have been ruthless, but he was never given the chance. His loving wife never suspected he had a mistress, despite all the nights he spent "in town," nor did his devoted mistress ever so much as threaten to Google his wife. Sure, occasionally there was the *possibility* of a little conflict. Bruno would be walking into an elaborate ambush replete with guns, knives, brass knuckles, and, in case all else failed, a car bomb. But just as he was reaching for the anthrax-coated doorknob, the rival thugs would get a call telling them everything had been resolved, so they'd yank the door open, give Bruno a bear hug, and then they'd all sit down for an espresso and some nice biscotti.

The author was a successful businessman who, at sixty-something, was still married to his childhood sweetheart and had several smart, well-adjusted kids. I asked him how he felt about conflict in his real life. He frowned. "I don't like it," he said, tensing. "Who does?"

The answer, of course, is no one (drama queens notwithstanding). That's exactly why we turn to story—to experience all the things that in life we avoid, rationalize away, fear, or long to accomplish but for various and sundry reasons haven't or can't. We want to know what it would cost us emotionally, what it would really feel like, should we ever find ourselves, or someone we know, in a similar situation. It boils down to this: in real life we want conflict to resolve right now, this very minute; in a story we want conflict to drag out, ratcheting ever upward, for as deliciously long as humanly possible.

But wait! I hear you asking: *If we're wired to feel what the protagonist does, since conflict hurts, are you saying we're masochists?* Not at all. In the same way that a vicarious thrill, being one crucial step removed, isn't nearly as powerful as the real thing, neither is the pain we experience when lost in a story. Sure, we're literally *feeling* what the protagonist feels, but our trusty brain is also quite aware that what befalls the poor sap is not, in fact, literally *happening* to us. So, although we feel Juliet's anguish on awaking to find Romeo lifeless by her side, we never once lose sight of the fact that our own beloved is, in fact, snoring peacefully in the theater seat next to us. And that, my friends, is what makes stories so deeply satisfying. We get to try on trouble, pretty much risk-free.

And so in literature, *not as in life*, the goal is to embrace conflict and harness it to suspense. Now, for the sixty-four-thousand-dollar question: how do you translate impending conflict into ongoing suspense?

Suspense Is the Handmaiden of Conflict

As we know, a story spans the distance from "before" to "after," when things are in flux. Therefore a story inherently chronicles something that is changing. Usually that "something" revolves around a problem the protagonist must solve in order to actually get from the shores of "before" to the banks of "after."

On the surface, conflict is borne along by escalating external obstacles that keep the protagonist from quickly solving the problem and getting on with his day, no worse for wear. But those obstacles mean nothing unless, *beneath* the surface, the seeds of that conflict are present from the outset, as they begin pushing their tender shoots through the soil in search of the sun. Picture it as the first hairline crack in the otherwise solid wall of "before." The cause of this fissure is often the answer to the question, *Why does this story have to start at this very minute?* For instance, in Anita Shreve's *The Pilot's Wife*, it's the ominous sound of a predawn knock at the door that tells Kathryn, the protagonist, that something is very wrong. This initial crack causes the plaster to slowly crumble as Kathryn discovers that her husband, Jack, was in many ways a complete stranger. The story, however, isn't in the facts of Kathryn's situation, per se, but in how she struggles to make sense of them, given that everything she believed to be true up to that moment, wasn't. Talk about the kind of change we're wired to resist. Everything in Kathryn wants to believe Jack was the perfect husband, if only life (aka the story) would stop poking holes in her carefully—and largely subconsciously—constructed rationalizations.

Thus a story's first hairline crack and its resulting offshoots are like fault lines, running through the center of the protagonist's world, undermining everything. As with an earthquake, the cracks tend to be caused by two opposing forces, with the protagonist caught between them. I like to think of these battling forces as "the versus," which taken together create the arena in which the story then proceeds to duke it out. Keeping in mind that every story has more than one versus, here are the most common:

- What the protagonist believes is true versus what is actually true

- What the protagonist wants versus what the protagonist actually has

- What the protagonist wants versus what's expected of her

- The protagonist versus herself

- The protagonist's inner goal versus the protagonist's external goal

- The protagonist's fear versus the protagonist's goal (external, internal, or both)

- The protagonist versus the antagonist

- The antagonist versus mercy (or the appearance thereof)

So, to enlarge our nutshell a bit: story takes place in the *time* between "before" and "after" and in the *space* between the "versus," as the protagonist maneuvers within two conflicting realities, trying to bring them into alignment (and thus solve the problem). Once he does this, the space between them closes and the story ends. Meanwhile, the suspense builds as the reader wonders how on earth these realities, which seem to be moving further and further apart, will *ever* come together.

In short, it's the story's job to poke at the protagonist, in one way or another, until she changes. With that in mind, let's take a look at how the "versus" can shape a story from the inside out.

The Tale of Rita and Marco: Versus by Versus

Before we dive into their story, let's review three important facts about how our brain processes info:

1. As we'll explore in chapter 10, the brain is wired to hunt for meaningful patterns in everything, the better to predict what will happen next based on the repetition or the alteration of the pattern (which means, first and foremost, that there need to be meaningful patterns for the reader to find).[8]

2. We run the scenario on the page through our own personal experience of similar events, whether real or imagined, to see whether it's believable (which gives us the ability to infer more information than is on the page—or go mad when there isn't enough information for us to infer anything at all).[9]

3. We're hardwired to love problem solving; when we figure something out, the brain releases an intoxicating rush of neuro-transmitters that say, "*Good job!*"[10] The pleasure of story is trying to figure out what's *really* going on (which means that stories that ignore the first two facts tend to offer the reader no pleasure at all).

All this is another way of saying the reader knows way more than you think she does, so relax and don't worry so much about giving too much away. Chances are your readers will be several steps ahead of your protagonist, which is exactly where you want them to be. For instance, the reader will have a much better handle on the likelihood of whether or not Marco, the office Lothario, will *actually* leave his wife the second she gets back from visiting her sick mother, than his fretful mistress Rita does—even though Rita is the first-person narrator. And that's a good thing. Because it means suspense arises not only from what we suspect the characters will do, but from the tension we feel watching Rita pick out her trousseau, knowing damn well that not only won't Marco be leaving his wife, but there's a good chance he doesn't even *have* a wife.

Thus while we're rooting for Rita, the last thing we're hoping for is that she'll actually land Marco, even though we're in her skin and can feel how strong her desire is. Instead we're hoping she'll realize Marco is actually the *last* thing she needs, before it's too late and, god forbid, she actually gets him. Rita's real struggle—the one the reader is following with bated breath, the one the story is about—is internal. In other words, the story revolves around how Rita *views* her world rather than what happens in it. Therefore there are myriad layers of conflict laced into Rita's story. Let's take them versus by versus, shall we?

On the external level we have what Rita *wants* (Marco) versus what Rita *has* (Marco's promises). On the internal plane, the conflict is between what Rita *believes* (that Marco is her soul mate) versus what is *actually true* (Marco is soulless). This means that on the page we're watching Rita try to woo and win Marco, as the writer slowly reveals that Marco is a very different sort from what Rita imagines. This gives the reader the space to anticipate how Rita will feel when she finds out and what she'll do as a result.

This brings us to one of the most potent versus of all, one that often defines the playing field: what Rita *wants* (Marco's unadulterated love) versus what is *expected* of her (Marco expects her to turn a blind eye to his cheating). This means that throughout the story Rita will be struggling with the fact that Marco seems to believe all she wants from life is to cater to his every whim, no questions asked. Knowing how weak this makes her look to her friends, chances are she'll be struggling to at least *appear* to meet their expectations, too. She's going to dump him, she swears; she just hasn't found the right moment.

This, in turn, triggers that little voice in the back of her head that's worried they might be right about Marco. But because she's totally besotted with him, she ignores her suspicions. Aha! Now Rita is also battling *herself*, which will be evident in her internal response to Marco's actions. That is, she'll rationalize. This means that often what she says, and what she's actually thinking, will be at odds. Talk about a great way to ratchet up the tension!

Which brings us to the question: *why* would Rita ignore something that's abundantly clear to the rest of us? What we're looking for is the *reason* Rita is so desperate to hook up with Marco, beyond the fact that he sends her pulse through the roof. Okay, let's say Rita's deeper motive is that she is afraid of being alone.

Fear? Could this be grounds for yet another source of conflict? Rita's goal versus her fear, perhaps? Not quite. After all, rather than keeping her from her goal, her fear is part of what's driving her *into* Marco's arms, since if she lands him, she'll never have to confront her dread of being alone. Not quite a versus, yet. But then, we haven't explored Rita's internal goal yet. As fate (that is, the author) would have it, Rita's internal goal is to be loved for who she is by a man who is truehearted. Sound like Marco? Nope. Definitely a conflict there, and one that reveals a nifty rule of thumb:

> One way to tell if what the protagonist wants in the beginning is her genuine goal is to ask yourself: will she have to face her biggest fear, and so resolve her inner issue, to achieve said goal? If the answer is no, then guess what—it's a false goal.

And you know what that means? That Rita's fear is, in fact, part of a very compelling versus: her fear versus her *genuine* goal, which is to be loved by a truehearted man. Thus, if she's going to remain true to herself, she will shun Marco, even though it means being alone. Being aware of all these layers allows the writer to use Rita's fear of being alone to shape her reaction to everything that happens to her. Thus her external decisions, internal monologues, and body language will in some way reflect her true motive, whether *she* is aware of it or not. We're not talking something as obvious as Rita thinking this:

Gee, Marco sure is a big fat jerk, but since I'd rather die than be alone, I better do everything he wants me to, even if these damn stilettos kill my feet.

Instead it's more like this:

As Marco and I walked into the courtyard, I saw my neighbor, Mabel, scurry into her apartment, quickly closing her door lest one of her cats slip out. How many does she have? Eight, nine? Yet she always looks so sad, as if she's afraid even her cats don't like her. *There but for the grace of god,* I thought, grateful for the weight of Marco's arm around my shoulders, even if it means I have to walk faster to keep pace with him, which isn't easy in stilettos.

The more the reader is aware of Rita's true motivation, the more we'll understand why a woman so otherwise smart and savvy would go after a Neanderthal like Marco, and the more we'll be rooting for her to fall for one of Mabel's fluffy little kittens instead.

Which brings us to the most obvious source of conflict: the *antagonist*—in this case, Marco. But let's not flatter him with too much attention, narcissist that he is, because we are much more interested in Rita. She is the sun in our universe, and everything revolves around her. So when it comes to Marco, what we care about is how he will affect Rita.

Because Marco personifies the escalating obstacle that Rita needs to overcome, it's important that he put up a really good fight. This is crucial, since the protagonist is only as strong as the antagonist forces her to be. Readers are sticklers when it comes to the "prove it" department; in this they're a lot like the citizens of Missouri, the "Show Me" state. They have no intention of taking anyone's word for how courageous the protagonist is. After all, anyone can say they're brave. Or daring. Or worthy. Does that prove anything? Only that he's a braggart, a bore, and, most likely, a coward. In fact, those who are truly brave tend to see themselves as not brave at all.

The point is, the antagonist must put the protagonist through her paces. This means Marco must do everything he can to rope Rita in, except, of course, actually become the man she hopes he is. Because what Rita needs far more than Marco is the ability to face her fear. Which means that by mercilessly leading her on, Marco is actually doing her a favor by forcing her to confront the thing that's always held her back. And that is precisely what the reader will be rooting for.

Most of the time.

Because there's that one last versus to contend with: The antagonist versus mercy (or the appearance thereof). No one is bad to the bone, psychopaths notwithstanding. And in the case of psychopaths, their defining trait is the ability to feign empathy without actually feeling a thing. Guys like serial killer Ted Bundy are utterly charming and appear capable of mercy, right up to the moment they break out the duct tape and the hacksaw. The key element in the "mercy" rule is the implied "maybe." *Maybe*, against all odds, Marco will change. You want there to be a moment or two when the reader thinks, *Hey, seems like Marco isn't so bad after all.* This moment will probably come just as Rita is in the midst of deciding never to see him again. So she relents. And for a minute it looks as if it'll turn out okay after all. And then, when he thinks no one is looking, Marco kicks one of Mabel's cats really hard and we think, *Uh-oh. . . .*

Why is this important? Because it's difficult to maintain suspense in the face of a foregone conclusion. Even a smidgeon of "maybe" goes a very long way. If your force of opposition—whether a femme fatale, a cad, or a cyborg—is all bad, why bother having them show up? All they have to do is phone in the threat. And with caller ID, who's going to answer, anyway? However, if the protagonist has a nasty case of the flu, and Ted Bundy shows up with a steaming bowl of homemade chicken soup, well, that's another story. Maybe he's had a change of heart. Or maybe the soup is laced with arsenic. The point is, we don't know. Hello, suspense!

The reason the various versus are so good at engendering suspense is that pitting two opposing desires, facts, or truths against each other inherently incites ongoing conflict. It gives the reader something to root for, another yardstick by which to measure the protagonist's progress, and a clear view of where the conflict lies. And so it might come as a surprise that writers often work overtime, devising ingenious plot twists to keep that very suspense under wraps. Which means it's time to "reveal" how one of the most popular methods writers employ to add suspense often produces the exact opposite effect.

MYTH: Withholding Information for the Big Reveal Is What Keeps Readers Hooked

---

REALITY: Withholding Information Very Often Robs the Story of What Really Hooks Readers

First, what is a reveal? A reveal is a fact that, when it finally comes to light, changes (and in so doing, explains) something—often, that something is "everything."

A major reveal is the surprise near the end that twists the meaning of everything that came before it. It's Darth Vader booming, "I am your *father*, Luke"; it's Evelyn Cross Mulwray admitting to Jake

Gittes, "She's my sister *and* my daughter"; it's Norman Bates in his dead mother's dress.

These reveals are shocking, yet they are completely believable the second we hear them. Why? Because up to that moment, although the story made sense, we couldn't quite shake the feeling that there was more going on than met the eye—which we actively tried to make sense of throughout. This is something we were able to do because the writers gave us a specific pattern of hints all the way along. And so, although each story made sense up to that moment, in light of the reveal it makes even *more* sense.

But make no mistake, it is *only* because of the pattern of hints that the reveal, when it comes, is instantly accepted as truth. Otherwise, it's one of the three dreaded Cs: a *convenience*, a *contrivance*, or a *coincidence*. It's like reading a murder mystery in which we find out on the last page that the hero, about to go to the gallows for a murder he didn't commit, has a guilty evil twin who no one (including, one suspects, the writer) knew existed up to that moment.

The problem with such books is that because the author has kept so much crucial information secret, we have *no* idea what is really going on, nor do we have a way of figuring it out. Or worse, we don't even know that there *is* anything going on beyond what's on the page. For instance, I once read a five-hundred-page manuscript about Fred, an unscrupulous automotive executive who staked his company's fortune on a car that, on the eve of its unveiling, he discovered had a potentially fatal design defect. Squelching the information, Fred put it on the market anyway, with the expected tragic results. The novel was about how Fred was then brought to justice. The manuscript contained no surprises until page four hundred and fifty. That's when it was revealed that Fred had been the subject of an ongoing undercover FBI investigation from the start. In fact, several of his close personal associates, including Sally, his mistress, had been spying on him from the get-go. There wasn't a whiff of this in the manuscript, mind you, not even the slightest teeny tiny hint. When I asked the author about it, he smiled and said he'd done it on *purpose*, because he was saving it for a big reveal at the end.

The trouble was, no one would ever have read that far. Why? Because by working mightily to keep the reader in the dark, he had robbed the story of what would have been its primary source of tension and suspense. Talk about irony! The truth, which is completely obvious in hindsight, is simply this:

> If we don't know there's intrigue afoot, then there is no intrigue afoot.

Because while readers relish looking back and reinterpreting specific events in light of new information that now twists their meaning, there are two ironclad conditions that must be met first:

1. There must have been a pattern of specific "hints" or "tells" along the way, alerting us that all was not as it seems, which the new twist now illuminates and explains.

2. These "hints" and "tells" need to stand out (and make sense) in their own right *before* the reveal.

What readers won't do is go back and insert entire subplots. It's like saying, "Hey, I know watching Fred for four hundred and fifty pages was dull, but now go back and reimagine the whole thing knowing that the FBI was always just outside his door, listening. And all those people who claimed to be his friends? They were secretly wired. And Sally, his mistress? She never even liked him."

To make matters worse, in light of this reveal, everything Fred's friends did back then no longer rings true. Because *had* they been wired, they would have been nervous and it would have shown, if only in their body language. There would have been something in Sally's behavior that intimated she was up to more than an afternoon's delight. Sure, the kindest among us might think, *Well, I guess since Sally was actually working for the feds, she's a pro, so there's no way she'd have done anything that would tip Fred off.* Trouble is, that *still* won't make the scene in which she was hiding her true feelings any more compelling or believable, given what we know about the infallibility of body language and our propensity for making inadvertent mistakes.

Not that we have to know (or even suspect) what Sally is *really* up to. But we do have to know that something about Sally's behavior is "off," thus alerting us to the fact that there's more going on than meets the eye. You *want* us to try to figure out what that might be. To that end, you *can* mislead (as opposed to lie to) us along the way. Take Hitchcock's *Vertigo*, in which retired police detective "Scottie" Ferguson is led to believe that a beautiful young woman named Madeleine is the troubled wife of his old friend Gavin Elster, who has hired Scottie to make sure she doesn't kill herself. As Scottie then falls in love with the enigmatic Madeleine, we sense both her attraction to him and her reluctance to surrender to it—giving these scenes tension and suspense. We chalk this up to the very believable fact that since she's not only married, but to a good friend of his, she feels doubly guilty—still, she doesn't quite seem as crazy as Elster intimated. So when we find out what was *really* going on—she *is* in love with Scottie, but *isn't* married to Elster, who hired her to set Scottie up—on reflection her behavior makes even *more* sense, totally validating the reveal.

Contrast that with the ballad of auto exec Fred and undercover agent Sally. Because the author steadfastly kept any hint of conflict out of their trysts, we had no idea there was more to it than what was on the page, so it was pretty dull. But not to the author, who knew Sally was hiding the truth from Fred, which no doubt made it very exciting indeed—to him. Why deny the reader that same pleasure?

An Irony: Reveals Often Obscure

When done properly, reveals can be extremely effective. But they're woefully overused and almost always to ill effect—perhaps because writers rarely seem to ask themselves this crucial question:

What does holding back this information *gain* me, story-wise? How does it make the story *better*?

I believe the misuse of reveals often stems from a fundamental misunderstanding, so let's begin there. Some writers, knowing it's crucial that the reader feel an instant sense of urgency that makes her want to know what happens next, believe that keeping something "secret" will do the trick. Then surely she'll read on to find out what the secret is, right? These writers tend to forget that first we have to *want* to know the secret (not to mention know that there *is* one). Therefore, the way to lure the reader in is definitely *not* by either

- Keeping the *real* reason the characters are doing what they're doing so secret that we don't even know there *is* a real reason

Or, even more common

- Letting us know there *is* a secret, but then keeping it so vague that we can't even *guess* what the particulars are

The trouble with both methods is they presuppose that we're already engaged enough to care what happens to the characters. Ironically, more often than not it's the very information the writer's withholding that would make us care. That's because the reveal usually requires couching the most interesting information in vague generalities, and as we saw in chapter 5, little good ever comes of that. So while we know the protagonist, Bob, has a "problem" that got him fired, the writer decides to withhold both what that problem is and what kind of work Bob does, the better to wow us later with the fact that Bob is *actually* a toy poodle who lost his job at Sea World because he found hopping across the stage on his hind legs demeaning, so he chased a squirrel instead. Now, that's sort of interesting. But because there are *so* many details the writer has to keep hidden or risk spoiling the reveal, for the first hundred pages or so we are allowed to think Bob is just an unusually hairy guy who's down on his luck, which is why he lives in a crate under the freeway. Thus the only thing we *are* clear about is that we don't really know what's going on.

The trouble with keeping both the situation and the characters generic—since anything else would "give it away"—is that it not only straitjackets the story but also tends to strip the characters of their credibility as well. Why? Because once the writer decides to keep the protagonist's big secret under wraps, the protagonist can't so much as *think* about it—even though, of course, it's exactly what he would be thinking about. Even more damaging, he can't react to anything the way he would, given what really happened, because that, too, would give it away. So when the reveal finally comes, nothing he's done up to that point is in any way consistent with what a person in that situation would have done. Thus the reveal becomes a groaner.

The good news is, there is another way.

The Beauty in Showing Your Hand

What if you lay your cards on the table face up? What difference does it make in terms of suspense? Let's try a little test.

First, a description of a scene with the cards held close to the vest: Val is looking for her roommate Enid, who's hours late. After canvassing the neighborhood, she reluctantly knocks on the door of her new neighbor, Homer, shows him a photo of Enid, and asks if he's seen her. He says no, but seeing how worried Val is, he invites her in for a soothing cup of herbal tea. Realizing she probably *is* blowing the whole thing out of proportion, and that Homer's really cute, she accepts. Over two steaming mugs, Homer reassures Val, suggesting that Enid probably just decided to visit a friend, nothing to worry about. Half an hour later Val leaves, feeling relieved and wondering whether Homer is single.

Now, imagine the exact same scene, except we've known the whole time that Homer has Enid locked in the basement, where she can hear the conversation upstairs and is trying desperately to get out. This time, we're riveted, rooting for Enid, and praying Homer hasn't slipped a roofie into Val's Sleepytime Tea.

But does that mean that you have to lay *all* your cards on the table? Can't you keep a few of them up your sleeve, for use later? Absolutely; there's nothing readers love more than to be fooled—as long as, once the truth is revealed, everything still makes complete sense, both in the moment it happens and in hindsight after the "real truth" is revealed.

So let's return to the saga of Val and Homer, with Enid handily duct-taped to a chair in the basement. This time, let's imagine that as Homer and Val talk upstairs, Enid manages to free herself, climb out the basement window, and run home. Now we're just dying for Val to get the hell out before Homer ties her up too. So when she finally leaves, we breathe a huge sigh of relief.

But as Homer's closing the door, his phone rings. It's his boss at the FBI. Reinforcements are on their way; she can't believe he captured the notorious hacksaw killer Enid Dinsmore a mere week after going undercover, especially given that they just received intelligence that Enid plans to kill again, tonight. Something about a roommate named Val.

Again, the tension is palpable, isn't it? Because conflict, as it turns out, isn't ephemeral at all. It's visceral. It's the space you leave for the reader, allowing her to leap into the fray and imagine the possibilities. Never forget that story unfolds in the space between two opposing forces. If you make sure the reader's always aware of the conflicting realities the protagonist finds herself trapped between, you'll be off to the races—together.

CHAPTER 7: CHECKPOINT

Have you made sure that the basis of future conflict is sprouting, beginning on page one? Can we glimpse avenues that will lead to conflict? Can we anticipate the problems that the protagonist might not yet be aware of?

Have you established the "versus" so that the reader is aware of the specific rock and hard place the protagonist is wedged between? Can we anticipate how he will have to change in order to get what he wants?

Does the conflict force the protagonist to take action, whether it's to rationalize it away or actually change? Imagine what *you* would want to avoid if you were your protagonist—and then make her face it.

Have you made sure that the story *gains* something by withholding specific facts for a big reveal later? Don't be afraid of giving too much away; you can always pare back later. Showing your hand is often a very good thing indeed.

Once the reveal is known, will everything that happened up to that point still make sense in light of this new information? Remember, the story must make complete sense without the reveal, but in light of the reveal, the story must make *even more* sense.

8

CAUSE AND EFFECT

COGNITIVE SECRET:

From birth, our brain's primary goal is to make causal connections—if this, then that.

STORY SECRET:

A story follows a cause-and-effect trajectory from start to finish.

People assume that the world has a causal texture—that its events can be explained by the world's very nature, rather than just being one damn thing after another.

—STEVEN PINKER, *How the Mind Works*

WE'RE OFTEN WARNED not to make assumptions. Did you ever actually *try* it? It's like saying, "Don't breathe." We make assumptions about everything, every second of the day—largely because, after breathing, our survival depends on it. We assume that if we cross the street without looking, we might be mowed down; we assume that eating the leftover creamed tuna we accidentally left on the counter overnight is probably not such a bright idea; we assume that if the phone rings after two a.m. it can't be good. If we couldn't assume the result of, well, anything, why would we chance getting out of bed in the morning? So we assume. As philosopher David Hume pointed out, as far as we're concerned, causality is "the cement of the universe."[1]

Are our assumptions sometimes wrong? Decidedly. Here's an apt case in point, from Antonio Damasio: "Usually the brain is assumed to be a passive recording medium, like film, onto which the characteristics of an object, as analyzed by sensory detectors, can be mapped faithfully. If the eye is the passive innocent camera, the brain is the passive, virgin celluloid. This is pure fiction." Instead, Damasio explains, "Our memories are *prejudiced*, in the full sense of the term, by our past history and beliefs."[2]

In other words, our assumptions are based on the consequences of our prior experiences. But we don't stop there. While a few other species

145

take a rudimentary stab at observing and predicting what might happen next, we alone try to explain why.[3] Understanding why "this" caused "that" is what allows us to anticipate what might happen next and decide what the hell we're going to do about it. It lets us theorize about the future and, better yet, try to change it to our advantage.

As for the wrong assumptions this sometimes begets? It's our acceptance of fallibility that makes us human, as evidenced by the courage we muster to take risks, knowing things might not turn out as planned. It's when they don't that people usually tell us not to make assumptions. What they really mean is, *the assumption you're making isn't working; try another.* Because very often, as Kathryn Schulz of *Being Wrong* attests: "We think this one thing is going to happen, and something else happens instead."[4]

Story arises from the conflict between "this one thing we thought was going to happen" and "what happened instead." It then plays out in a clear cause-and-effect trajectory from start to finish—otherwise, it would be "just one damn thing after another." So in this chapter we'll determine how to make sure your story follows one astonishingly simple mantra; we'll explore how to harness the external cause and effect of your plot to the more powerful internal cause and effect of your story; we'll take a look at why "show, don't tell" is a matter of *why* rather than *what*; and we'll introduce the all-important "And so?" test to guarantee that neither cause-and-effect trajectory ever goes off the rails.

The Logic of If, Then, Therefore

As we know, both life and story are driven by emotion, but what they're ordered by is logic. Logic is the yang to emotion's yin. It's no surprise that our memories—how we make sense of the world—are logically interrelated. According to Damasio, the brain tends to organize the profusion of input and memories, "much like a film editor would, by giving it some kind of coherent narrative structure in which certain actions are said to cause certain effects."[5]

Since the brain analyzes everything in terms of cause and effect, when a story *doesn't* follow a clear cause-and-effect trajectory, the brain doesn't know what to make of it—which can trigger a sensation of physical distress,[6] not to mention the desire to pitch the book out the window. The good news is, when it comes to keeping your story on track, it boils down to the mantra *if, then, therefore. If* I put my hand in the fire (action), *then* I'll get burned (reaction). *Therefore*, I'd better not put my hand in the fire (decision).

Action, reaction, decision—it's what drives a story forward. From beginning to end, a story must follow a cause-and-effect trajectory so when your protagonist finally tackles her ultimate goal, the path that led her there not only is clear, but, in hindsight, reveals exactly why this confrontation was inevitable from the very start. Note the critical words *in hindsight*. Everything in a story should indeed be utterly predictable, but only from the satisfying perspective of "the end."

This is not to say that a story has to be linear or that the cause-and-effect route it takes must be chronological—quite the contrary. It can take death-defying leaps in time and location and even be told backward: witness Martin Amis's novel *Time's Arrow*, Harold Pinter's play *Betrayal*, and Christopher Nolan's film *Memento*. But what must move forward from page one is the clear logic of the emotional arc—that is, the story the reader is following. Even a book as seemingly "experimental" as Jennifer Egan's Pulitzer Prize–winning *A Visit from the Goon Squad*, which is told in standalone short stories that follow several characters back and forth in time, follows a novel's arc. Egan herself says, "When I hear that something is experimental, I tend to think that means the experiment will drown out the story. If having a story that's compelling—[that makes] you want to know what will happen—is traditional, then ultimately I am a traditionalist. That is what readers care about. It's what I care about as a reader."[7]

To create a story the reader will care about, the narrative must follow an emotional cause-and-effect trajectory from the outset. How? By obeying the basic laws of the physical universe. Thus the key thing to remember is, naturally, Newton's first law of thermodynamics: you can't get something from nothing. Or as the equally brainy Albert Einstein reportedly quipped, "Nothing happens until something moves." In other words, no matter how much something catches you off guard, nothing ever really occurs out of the blue. Not in real life, not in a story. There is always a cause-and-effect trajectory, whether or not the protagonist—or in the case of real life, you and I—see it coming.

We tend to be blissfully clueless as to the fly ball that is about to bean us—a ball everyone else has been watching since the batter smacked it up into the air. So although Leslie has no idea that her boyfriend Seth is sleeping with Heidi in accounting, the whole office figured it out the instant he began misting up over the magnificence of Heidi's spreadsheets. Hence, although when Leslie finally finds out that Seth's a big fat cheater, it'll be news to her, her coworkers will have spent weeks taking bets on who she'll take down first—Seth or Heidi. Of course, once Leslie finds out about their little tryst, she's going do some looking back herself, and damned if she won't discover that there is, in fact, a pattern of telltale signs, which she'll now see with dizzying clarity, as if they were dominoes neatly lined up to fall just so.

The one difference, however, between the way Newton's law works in a story and the way it works in real life is that in real life there will be a million irrelevant things happening at the same time, whereas in a story *there will be nothing that does not in some way affect the cause-and-effect trajectory*. It's the writer's job to zero in on the story's particular "if, then, therefore" pattern and stick with it throughout. This trajectory is the track that the story's narrative train rumbles down. Sure, it might have twisty hairpin curves, switchbacks, harrowing ups and downs, even a few reversals, but the train itself never jumps track, derails or, hopefully, runs out of steam.

But wait—with all due respect to Jennifer Egan, what *about* experimental literature? What about avant-garde fiction? It doesn't seem as bound by the laws of cause and effect, or by any laws at all, for that matter. In fact, some say its *raison d'être* is to prove that fiction doesn't need a plot, a protagonist, characters, internal logic, or even actual events, to show that we've risen above all that. And hey, what about *Ulysses*, the first novel to leap headfirst into that seductively self-reflective stream of consciousness? Isn't it widely acclaimed as the best novel ever written? That was a pretty experimental book in its time. Let's dive into this question a wee bit deeper, shall we?

MYTH: Experimental Literature Can Break All the Rules of Storytelling with Impunity—In Fact, It's High Art and Thus Far Superior to Regular Old Novels

REALITY: Novels That Are Hard to Read Aren't Read

A few years back, Roddy Doyle, widely regarded as Ireland's best contemporary novelist, stunned an audience gathered in New York to celebrate James Joyce by saying "*Ulysses* could have done with a good editor." Warming to his topic, he went on to muse, "You know, people are always putting *Ulysses* in the top ten books ever written, but I doubt that any of those people were really moved by it."[8]

People like to tackle *Ulysses* in part because it's such a hard read that making it to "the end" is a testament to their intelligence (if not endurance). But no matter how smart they are, few people actually enjoy reading it. The trouble is, even unread, such books can do great harm. According to author Jonathan Franzen, books like *Ulysses* "send this message to the common reader: Literature is horribly hard to read. And this message to the aspiring young writer: Extreme difficulty is the way to earn respect."[9] And therein lies the real problem.

There is a school of writing that holds that it's the reader's responsibility to "get it," rather than the author's job to communicate it. Many writers of experimental fiction graduated from this particular school with advanced degrees. Thus, when we readers don't "get it," the fault is not assumed to be theirs, but ours. This attitude can foster an unconscious contempt for the reader, while freeing the writer from any responsibility beyond his or her own self-expression. It also tends to presuppose the reader's interest and earnest dedication from word one—as if somehow the reader owes it to the author to choke down every single word.

The trouble is, reading novels freed from the supposed plebeian constraints of plot, character, and even a nodding acquaintance with cause and effect, quickly becomes work. But unlike most of the work we willingly undertake—like heading into the office every day, weeding the garden, or housebreaking that cuddly new pup—it can be hard to see what reward slogging through to the bitter end will bestow. That is, unless reading a book meant to bore you so as to give you the experience of being bored sounds riveting (which, of course, would defeat the purpose). Much more common is the experience a student recently shared with me. She'd just gotten an MFA from a very prestigious university and confessed that many of the books she was required to read made her cry—because they were so mind-numbingly boring. Probably not the author's intention.

But there's a deeper, and more interesting, question in play here. Given that story is a form of communication, one we're wired to respond to, what are these novels, anyway? Are they stories at all? In many cases, the answer is a resounding no. This isn't to say that a select group of readers might not learn something from them—after all, we learn from textbooks, math equations, and dissertations. And there can pleasure in them too. But the pleasure doesn't come from the joy of reading a compelling story as much as from having solved a difficult problem, which is genuinely intoxicating. It makes you feel smart, like doing the Sunday crossword puzzle in ink. There's nothing wrong with that.

What is wrong, though, is the notion that if you're really enjoying a story, it automatically means that both you and the story are woefully lowbrow. The irony is that the hardwired pleasure a good story brings proves it's necessary to our survival. Just as we evolved biologically to find food tasty so we'd eat it, story triggers pleasure so we'll pay attention to it.

As writer A. S. Byatt so eloquently says, "Narration is as much a part of human nature as breath and the circulation of the blood. Modernist literature tried to do away with storytelling, which it thought vulgar, replacing it with flashbacks, epiphanies, streams of consciousness. But storytelling is intrinsic to biological time, which we cannot escape."[10]

And who would want to? The good news is that experimental fiction can be harnessed to what the reader is wired to respond to. In fact, the best of it already is. Which brings us right back to Jennifer Egan—who, having avowed that the most important thing is that the reader wants to know what happens next, adds, "Now if I can have that along with a strong girding of ideas and some kind of exciting technical forays—then that is just the jackpot."[11]

Hitting the jackpot means finding the narrative thread that gives meaning to everything that happens in your glorious experiment. So let's get back to figuring out exactly how to do just that.

The Two Levels of Cause and Effect

Whether experimental, traditional, or somewhere in between, we know a story plays out on two levels at once—the protagonist's internal struggle (what the story is actually about) and the external events (the plot)—so it's no surprise that cause and effect governs both, allowing them to dovetail and thus create a seamless narrative thread.

1. *Plot-wise* cause and effect plays out on the surface level, as one event logistically triggers the next: Joe pops Clyde's shiny red balloon; Joe gets kicked out of clown school.

2. *Story-wise* cause and effect plays out on a deeper level—that of meaning. It explains *why* Joe pops Clyde's balloon, even though he knows it will probably get him expelled.

Since stories are about how what happens affects someone—Joe, for instance—the reason he popped the balloon is more important than the fact that he popped it. In short, the *why* carries more weight than the *what*. Think of it as a pecking order: the *why* comes first, because it drives the *what*; the *why* is the cause; the *what* is the effect. Let's say, for example, Joe knew Clyde is secretly a killer clown and was about to use the balloon to lure a trusting tot into the deserted big top. Although Joe's dream has always been to pile into that teeny tiny car with all the other clowns, he knows he will never be able to live with himself if he doesn't stop Clyde, so he pops the balloon. Thus the *story-wise* cause and effect is not about *how* your protagonist gets from point A (being in clown school) to point B (not being in clown school), it's about *why*. The internal, story-level cause-and-effect trajectory tracks the evolution of the protagonist's inner issue, which is what motivates his actions. It reveals how he makes sense of what happens in light of his goal, and how he arrives at the decision that catapults him into the next scene.

It may come as a surprise that this is, in fact, what's meant by that perennial old saw, "Show, don't tell," which might just be the most woefully misunderstood writing maxim on the books.

MYTH: "Show, Don't Tell" Is Literal— Don't Tell Me John Is Sad, Show Him Crying

REALITY: "Show, Don't Tell" Is Figurative— Don't Tell Me John Is Sad, Show Me *Why* He's Sad

If there's one thing writers are told from the get-go, it's "show, don't tell." Good advice. Trouble is, it's rarely explained, so it's often completely misconstrued by being taken *literally*, as if "show" inherently means visually, from the outside in, as if you were watching a film. So when a writer hears, "Don't tell me that John is sad, *show* me," she spends hours writing how "John's tears fell like a torrential rainstorm, flooding the basement in a glittering release of everything he'd held in for so very long, knocking out the power and nearly drowning the cat." No, no, no! We don't want to see John cry (the effect); we want to see *what made him cry* (the cause).

What "show" almost always means is, *let's see the event itself unfold.* Instead of *telling* us that when John's father unexpectedly booted him out of the family business in front of everyone at the yearly stockholder's meeting, he cried a river, *show* us the scene in which he was ousted. Why? There are two very good reasons:

1. If you tell us after the fact that John was fired, it's a done deal, so there's nothing to anticipate. Worse, it's opaque—meaning there is nothing we can learn from it because we don't even know what, exactly, happened. But if you *show* us a scene in which John strides into the board meeting, sure he's going to be made CEO, well then, anything could happen (hello, suspense!)—*and we'd get to see it.* He could talk, blackmail, or yodel his way back into his dad's good graces, or he could surprise everyone by quitting first—which would mean that those tears we watched him cry were tears of *joy.* Scenes (even flashbacks) are immediate and fraught with the possibility that all could be lost—or gained. The same info, summarized after the fact? It's yesterday's news.

2. If we watch the stockholder's meeting, chances are we'll learn *why* John was fired, what John's father actually said, and how John reacted in the moment, which will give us fresh insight into their dynamic and who they are when the chips are down. This is where a lot of those missing specifics we were talking about in chapter 6 tend to be hidden.

In short, "telling" tends to refer to conclusions drawn from information we aren't privy to; "showing," to how the characters arrived at those conclusions in the first place. Thus "show, don't tell" often means *show us a character's train of thought*. I once worked with a writer whose protagonist, Brian, had a habit of swearing he'd never do something and then, for no apparent reason, doing it. Because this completely undermined Brian's credibility, I advised the writer to *show* him making each decision. What I got back was a manuscript full of passages like this:

> "Please, Brian dear, I know you said you'd never ever have a dog again, after what happened to Rover, but I saw the cutest cockapoo at the pound. What do you say?"
>
> Brian sat on the couch, staring pensively out the window, stroking his chin. Seconds ticked by. Finally, heaving a sigh, he said, "Okay, honey, let's go to the pound."

It wasn't until I read the sixth or seventh such passage that it dawned on me the writer had indeed taken my advice. Sure enough, he was "showing" Brian making decisions. Which, of course, was not at all what I meant. I was talking about Brian's train of thought, the reasoning that led him to change his mind. Very often "show, don't tell" refers to the progression of the character's inner logic. As in, don't *tell* me Brian changed his mind; *show* me how he arrived at the decision.

So, does "show, don't tell" *ever* refer to showing something physical? It absolutely does, primarily in two instances:

1. **When we already know the "why":** After a harrowing scene in which Brenda cruelly breaks it off with an unsuspecting Newman, the writer would definitely want to swap, "Newman was sad," for a visual image that telegraphs his sorrow. It might be his tears, it might be the way his voice catches, it might be in the slump of his shoulders, it might even be the way he's curled up

on the floor in the fetal position, whimpering. But, and this is crucial, whatever Newman does must also tell us *something we don't already know*. Maybe we're shocked that a big strapping guy like Newman would cry at all; he must be more sensitive than we thought. Or perhaps Newman had been pretending that he didn't really care, so when we catch sight of his slumped shoulders, we realize he does.

2. **When the subject at hand is *purely* visual:** As Chekhov so famously said, "Don't tell me the moon is shining; show me the glint of light on broken glass."[12] However, I'd venture to say that if there *is* a glint of light on broken glass, that broken glass had better be there for a story reason. Either literally, because someone is about to step on it, or metaphorically, as in, Brenda's announcement is about to cut Newman to ribbons.

The "Put Your Money Where Your Mouth Is" Test

Since story, both internally and externally, revolves around whether the protagonist achieves his goal, each turn of the cause-and-effect wheel, large and small, must bring him closer to the answer. How? By relentlessly winnowing away everything that stands in his way—legitimate reasons and far-fetched rationalizations alike—until the clock runs down to "now or never." It's sort of like musical chairs, except each chair, and the reason it's yanked out of play, is unique. There's a method to the madness, because each cause-and-effect pairing specifically—and logically—spurs the next. Each scene's decision point is tested by the next scene's action. In other words, each scene makes the next scene inevitable.

Think of it as the "put your money where your mouth is" test. Every time the protagonist makes a decision, saying to herself, *Yep, this is the right choice, and here's why . . .* the story then sits back with a great big grin and says, *Oh yeah? Prove it.*

Here's an example most of us can relate to: It's Thanksgiving, and once again you overate, big time. As you peel off your formerly loose clothes, still stuffed to the gills and feeling a wee bit sick, you vow to forgo so much as a single bite of the leftovers tomorrow—who says you don't have will power? You feel pretty confident about your ability to reach this goal, especially since, at the moment, even the thought of food makes you queasy. And there you have it: action, reaction, decision.

Cut to the next morning. Your plan works really, really well—for a while. And then, as if the possibility never crossed your mind, you get hungry. Thus today's action puts last night's decision to the test. What do you do? If you're anything like me, you tell yourself that being a little chubby is your way of resisting restrictive societal norms, and you chow down with gusto until even your elastic-waist fat pants are tight. Which leads you to the decision that first thing tomorrow, you're going to find out what a lap band procedure is and whether or not your insurance covers it. Which ups the ante nicely, don't you think?

Maximizing the Catapult

To guarantee that the stakes ratchet ever upward, you want to make sure you've infused each cause with enough firepower to trigger an effect that packs an unexpected, yet perfectly logical, wallop. For instance, in the movie *The Graduate*, the last thing Benjamin Braddock wants to do is date Mrs. Robinson's daughter, Elaine. So when his parents force the issue, he comes up with a plan: he'll take Elaine out and be such a cad she'll never want to see him again. Problem solved. Confidently armed with this decision, he takes action. His plan works perfectly, except for one rather large, unexpected wallop: he falls for Elaine, whom he has now completely alienated. In short, by solving one problem he's created an even bigger one. Which catapults him into a new decision: find a way to win Elaine's love (without dwelling too much on what he'll say, should she ever ask about how he lost his virginity).

In the same way, your goal is to be sure each individual scene effectively uses its specific "action, reaction, decision" to evoke maximum tension and to up the odds. At the beginning of the scene, it helps to ask yourself, *What does my protagonist want to have happen during this scene?* That established, ask yourself, "What is at stake here?" What will it cost her to get what she wants? Armed with this info, you're ready to write the scene. When you finish it, before diving into the next scene, ask yourself these questions:

- Has the protagonist changed? He should start out feeling one way and end up feeling another—often the exact opposite of how he originally felt.

- Having weighed his options, given what was at stake, and then made a decision, does he now see things differently than he did when the scene opened?

- Do we know why he made the decision he did? Do we understand how he arrived at that particular conclusion, even—make that *especially*—if his reasoning is flawed? Can we see how this changed his assessment of what's going on and how he's adjusted his game plan accordingly?

Notice that once again it's the protagonist's *internal* reaction to what happens that not only dictates what happens next but also gives it meaning. I hit on this so hard because this tends to be a spot where many stories go irrevocably south. Something happens, but we have no idea how it affects the protagonist or what he makes of it, thus it has no emotional impact—and so no firepower. Since the reason for his external reaction is therefore opaque—we have no idea what it is or how he arrived at it—although things definitely *happen*, the story itself coasts to a standstill.

Cause and Effect Doesn't Mean Predictable

Lest you think all this will make your story hopelessly unsurprising, take heart. Although your story's course may be fated from the moment that first domino teeters and falls, that doesn't mean it's predictable. Mastering the relationship between internal and external cause and effect allows you to tease the reader in ways she will actually enjoy. Here are four areas of delicious unpredictability:

1. A clear cause-and-effect pattern is what allows us to focus on the story's continual wild card: what the protagonist will *actually* do, given what he has to overcome. Remember the power of the "versus"? There are always competing desires, fears, and thus, choices. Just like in life, nothing's easy.

2. There's an *appearance* of free will. Just because someone might do something, it doesn't mean she will. There are lots of different reactions, and subsequent decisions, that a particular action might evoke—even though in the end, when all is revealed, said reactions and decisions will, in retrospect, be the *only* ones the character could have made. In short, what looks like free will going in turns out to be fate, when looking back.

3. Just like the rest of us, characters are famous for utterly misreading signs and rushing headlong in the absolute wrong direction (witness just about any episode of the classic TV series *I Love Lucy*).

4. Remember those cards that writers love to keep up their sleeves? Strategically revealed new information can change how the protagonist interprets everything that's happened up to then, not to mention change how the reader interprets the protagonist's motives from that point on.

Note that because story is driven by the protagonist's internal struggle, all these possible plot twists tend to stem from her attempt to get the most by giving up the least. And just like in real life, as we'll explore in depth in the next chapter, this usually only succeeds in making the situation worse. After all, there are so many really creative ways the protagonist can shoot herself in the foot. The goal is to establish her motivation beforehand so that the instant she pulls the trigger, readers will be both surprised and sheepish. *Of course!* we say to ourselves. *I should have seen it coming.*

What happens when a story *isn't* governed by the laws of cause and effect? The answer is quite sobering.

The Consequences of a Cause Without an Effect

Let's say the writer needs to get his protagonist, Barbara, out of an unexpected tight spot. So he devises a scenario that saves her in the moment and then promptly forgets all about it without stopping to consider that introducing a fact or character to solve a problem in one scene generates ongoing expectations in the reader's mind *that never go away*. As we'll discuss in detail in chapter 10, we're wired to predict what will happen next, and the way we do this is by charting patterns. Familiar patterns are safe. Deviate from a pattern, and bingo, like the robot in *Lost in Space*, it's "Danger, Will Robinson!" and you have our attention. The deviation then becomes the lens through which we filter the action.[13]

For instance, let's say Ronald, Barbara's condescending womanizer of a boss, insists on driving her home after they've worked into the wee hours. Her heart sinks, but she accepts—she needs the job. She breathes a sigh of relief as he pulls into her driveway, but when he quickly hops down from his huge black SUV and trots around to open her door, she knows she's in trouble. Noticing his lecherous grin, she assures him that she can see herself in. But Ronald stands firm; he wouldn't dream of leaving a defenseless woman alone until he's absolutely sure her home is intruder free. With that, he slips his arm snugly around her waist, and Barbara knows she'd better act fast or she'll be in big trouble.

This means the writer now has to get Barbara out of the situation, without offending Ronald. So he has Barbara turn to Ronald with a sufficiently demure smile and purr, "Not to worry, I'm packing. Sure, I may be a bit rusty since my stint as a sharpshooter in special ops back in '06, but I'm pretty sure I can still hit a moving target dead on anywhere from, say, right about where you're standing to a half mile in any direction." And with that she purposefully reaches into her purse. Not waiting to see whether she's going for a snubnose .38 or her house keys, Ronald dashes back to the safety of his massive Hummer, climbs inside, and screeches away. Problem solved.

Except that from there on out, the reader will be actively wondering when something is going to happen that will make Barbara have to either whip out a gun and save the day or fess up that she doesn't actually know what special ops is; it just sounded like something she read once in a Tom Clancy novel. But the damage doesn't stop there. The fact is, raising expectations that have nothing to do with the story we're reading changes the way we interpret absolutely *everything* that happens from that moment on.

Let's say Babs's story is meant to be a lighthearted chick lit romance in which her biggest problem is convincing Kyle, the idealistic young doctor she is trying to land, that she is not having an affair with her sleazy boss, Ronald. Trouble is, the minute she mentions her special ops background, we're in an entirely different story. One far less lighthearted. Since a huge part of the pleasure of reading is trying to figure out what's going to happen, readers will be spinning possible scenarios all on their own—and you want those scenarios to relate to the actual story you're telling. The last thing you want the reader wondering is, *Gee, if Babs really is a member of special ops, then why is she working as a receptionist in a fertilizer factory in Des Moines for a low-class creep like Ronald? Hmmm, don't they make bombs out of fertilizer, and isn't her boyfriend Kyle just a little too cagey about his past? Sure, he says he worked for Doctors Without Borders, but who's to say he didn't do a little drug running on the side? Could it be . . .* and with that, the reader is off and running in a story that the writer never in his wildest dreams envisioned.

Each thing you add to your story is like a drop of paint falling into a bowl of clear water. It spreads and colors *everything*. As with life, new information causes us to reevaluate the meaning and emotional weight of all that preceded it, and to see the future with fresh eyes.[14] In a story, it influences how we interpret every single thing that happens—how we read every nuance—and in so doing raises specific expectations about what might occur in the future. Since what makes stories so compelling is the thrill of actually making these connections (we're all dopamine addicts!), the connections must actually *be* there. When they're not—when the writer inadvertently plants a piece of information that has nothing whatsoever to do with the narrative itself—the story in the reader's mind veers in an entirely different direction than where the story is actually headed. So although the author may have completely forgotten about Babs's supposed special ops training the minute Ronald drove away, the reader won't. This is exactly the sort of situation that prompted Chekhov to note to S. Shchukin, "If you say in the first chapter that there is a rifle hanging on the wall, in the second or third chapter it absolutely must go off. If it's not going to be fired, it shouldn't be hanging there."[15]

It's Like Math—But in a Good Way

The cause-and-effect imperative can feel daunting to a writer. How *do* you keep track of everything? How can you be sure that you aren't accidentally leading the reader astray? Since, as Harvard psychology professor Daniel Gilbert says, "every action has a cause and a consequence,"[16] perhaps we can transform it into a good old-fashioned, genuinely simple math test.

But first, let's recap what we already know about the laws of cause and effect as they apply to story. To wit, every scene must

- In some way be caused by the "decision" made in the scene that preceded it

- Move the story forward via the characters' reaction to what is happening

- Make the scene that follows it inevitable

- Provide insight into the characters that enables us to grasp the motive behind their actions

This means you can gauge whether a particular scene is part of the great chain of cause and effect by asking yourself these questions:

- Does this scene impart a crucial piece of information, without which some future scene won't make sense?

- Does it have a clear cause the reader can see (even if the "real reason" it happened will be revealed later)?

- Does it provide insight into why the characters acted as they did?

- Does it raise the reader's expectation of specific, imminent action?

Now, for the math test: when evaluating the relevance of each scene in your story, ask yourself, *If I cut it out, would anything that happens afterward change?* To paraphrase the late Johnny Cochrane, "If the answer's no, it's got to go." Hey, I didn't say it was easy—but neither is pouring your heart and soul into a story only to have it waylaid by a couple of sweet-talking digressions.

Why Digressions Are Deadly

Think back to the last time you read a novel that had you hooked. Remember the sensation in your stomach as you turned page after page, anxious to find out what happens next? That's the feeling of momentum, and it's visceral—it's your brain's way of keeping you hooked, the better to crib info that might come in handy later.

Okay, now imagine the story is a car and it's zooming ahead at sixty miles an hour. You've completely surrendered to its momentum; you're one with the story. Then a real nice field of flowers off to the left catches the writer's eye. So he slams on the brakes, and you slam your head against the windshield as he hops out and frolics in the meadow. Just for a lovely, lyrical second. Then he's ready to get back on the road. But will the story still be going sixty? No, because he just brought it to a dead stop, which means—provided he can coax you back into it—the story is now going zero. There's a good chance it won't ever get back up to speed, especially since you don't quite trust the writer anymore. He stopped the story once for no reason at all; who's to say he won't do it again? Plus, since the digression broke the chain of cause and effect, you aren't exactly sure what's going on anymore. In fact, you're probably still trying to figure out how frolicking in the meadow fits into the story, which of course it doesn't. This means you're now paying less attention to what's actually happening on the page, so you might miss the very thing that would otherwise get the story back on track.

And that, my friends, is why, when it comes to digressions, heartless as it may seem, you have to kill them before they kill your story. I suspect this is what Mark Twain meant when he said, "A successful book is not made of what is in it, but what is left out."[17]

It pays to remain hypervigilant, because digressions come in all shapes and sizes. They can be misplaced flashbacks, they can be subplots that have nothing to do with the story itself, and they can be itty bitty. A digression is any piece of information that we don't need and therefore don't know what to do with.

Arm yourself with the knowledge that everything in a story must be there for a story reason; it must be something that, given the cause-and-effect trajectory, the reader needs to know, *at that moment.* Thus there is a question you must ruthlessly ask about every last scrap in your story: "And *so?*"

Because if you don't ask it, the reader will.

The "And *So?*" Test

When you ask "And *so?*" you're testing for story relevance. What does this piece of information tell us that we need to know? What's the *point*? How does it further the story? What consequence does it lead to? If you can answer these questions, great. But often the answer is "Um, it doesn't."

For instance, imagine if in *It's a Wonderful Life* there was suddenly a scene in which George Bailey learns to fly fish. You'd scratch your head, thinking, "And I need to know that—*why?*" Perhaps you'd even wonder if it was meant as a metaphor—something about the old "Teach a man to fish and he can feed himself forever" parable, maybe? And while you debated this, chances are you'd miss the bit where Uncle Billy absently wraps the eight grand in his newspaper and accidentally tosses it onto Potter's lap, so for a long time after that, *nothing* would make sense. Thus, even though George might have had a great time fly fishing, we do not need to know about it. The fly fishing scene fails the "And *so?*" test, which is no doubt why Frank Capra wisely kept it to himself.

What about your story? Does it sometimes toddle off in interesting yet irrelevant directions likely to thwart the readers' hardwired need to sense—if not see—the causal connections? Why not break out the red pen and have at it? Don't be shy. You might want to keep Samuel

Johnson's advice to writers tucked in the back of your mind as you slash and burn: "Read over your compositions, and wherever you meet with a passage which you think is particularly fine, strike it out."[18]

CHAPTER 8: CHECKPOINT

Does your story follow a cause-and-effect trajectory beginning on page one, so that each scene is triggered by the one that preceded it? It's like setting up a line of dominoes—you tap that first one, and they all fall in perfect order as each scene puts the "decision" made in the prior scene to the test.

Does everything in your story's cause-and-effect trajectory revolve around the protagonist's quest (the story question)? If it doesn't, get rid of it. It's that easy.

Are your story's external events (the plot) spurred by the protagonist's evolving internal cause-and-effect trajectory? We don't care about a hurricane, a stock market crash, or aliens taking over planet Earth *unless* it somehow directly affects your protagonist's quest.

When your protagonist makes a decision, is it always clear how she arrived at it, especially when she's changing her mind about something? Don't forget, just because *you* know what your protagonist is thinking doesn't mean your readers will.

Does each scene follow the action, reaction, decision pattern? It's like the one, two, three of a waltz. Get that rhythm stuck in your head—action, reaction, decision—and then use it to build momentum.

Can you answer the "And *so*?" to everything in the story? Ask this question relentlessly, like a four-year-old, and the minute you can't answer, know that you're likely in the company of a darling, a digression, or something else likely to cause your story to go into free fall.

WHAT CAN GO WRONG, MUST GO WRONG— AND THEN SOME

COGNITIVE SECRET:

The brain uses stories to simulate how we might navigate difficult situations in the future

---

STORY SECRET:

A story's job is to put the protagonist through tests that, even in her wildest dreams, she doesn't think she can pass.

No one would ever have crossed the ocean if he could have gotten off the ship in a storm.

—CHARLES KETTERING

THERE'S AN OLD SAYING: good judgment comes from experience; experience comes from bad judgment. The trouble is, bad judgment can be deadly. It can lead you to ignore that funny squeak every time you hit the brakes, put off checking out that odd-shaped mole on your big toe, decide to invest every penny with that clever guy whose hedge fund always turns a hefty profit. Even worse, bad judgment can derail your social life—which is a much bigger deal than we often realize. As neuroscientist Richard Restak says, "We are social creatures, the need to belong is as basic to our survival as our need for food and oxygen."[1] So, since there are countless tricky situations in which good judgment comes in awfully handy, often the best—not to mention safest—experience to learn from is someone else's. Could this be where story came from?

It's a question neuroscientists, cognitive scientists, and evolutionary biologists spend a lot of time pondering: considering that the brain is always working overtime to figure out what's safe and what isn't, why would it permit us to put the oft-sneaky "real world" on hold and get lost in a story?[2] The brain never does anything it doesn't have to, so as neuroscientist Michael Gazzaniga notes, the fact that "there seems to be a reward system that allows us to enjoy good fiction, implies that there is a *benefit* to the fictional experience."[3]

What is the benefit, survival-wise, that led to the neural rush of enjoyment a good story unleashes, effectively disconnecting us from the otherwise incessant Sturm und Drang of daily life? The answer is clear: it lets us sit back and vicariously experience someone else suffering the slings and arrows of outrageous fortune, the better to learn how to dodge those darts should they ever be aimed at us.

As Steven Pinker says, in a story, "The author places a fictitious character in a hypothetical situation in an otherwise real world where ordinary facts and laws hold, and allows the reader to explore the consequences."[4] Since we're wired to feel what the protagonist feels as if it were happening to us, when it comes to experience, this is as close as we're going to get to having our cake and eating it too. Which, of course, is precisely the point.

This means the protagonist is a guinea pig, and whether we like it or not, guinea pigs suffer so we don't have to. But although guinea pigs have PETA to champion their rights, protagonists are on their own—and trouble really is their middle name. "For example," cognitive psychologists Keith Oatley and Raymond Mar write, "a difficult breakup between a literary protagonist and his or her beloved cannot help but lead us to explore what it would be like were we in the same position. This knowledge is an asset when the time comes for us to cope with such an event in our own lives."[5]

The catch is, your protagonist really truly does have to suffer—otherwise not only will she have nothing to teach us, but we won't have much reason to care about what happens to her, either. Like everything in life, this is much easier said than done. That's why in this chapter we'll explore why you're actually doing your protagonist a favor by setting her up for a fall (or three or four); why in literary fiction, the protagonist must suffer even more than in a commercial potboiler; how to make sure your protagonist's trouble builds; and why some writers find it impossible to be mean to their protagonist. Finally, we'll go through eleven devious ways to undermine your characters' best-laid plans.

No Pain, No Gain

Have you ever suspected that maybe, just maybe, in some small, relatively inconsequential way, you might be just a tiny bit of a sadist? Good. Because as much as you love your protagonist, your goal is to craft a plot that forces her to confront head-on just about everything she's spent her entire life avoiding. You have to make sure the harder she tries, the harder it gets. Her good deeds will rarely go unpunished. Sure, every now and then it'll seem like everything's okay, but that's only because you're setting her up for an even bigger fall. You want her to relax and let her guard down a little, the better to wallop her when she least expects it. You never want to give her the benefit of the doubt, regardless of how much you feel she's earned it. Because if you do, the one thing she won't earn is her status as a hero.

The irony is, you aren't being a sadist at all. You're doing it for her own good, because you want her to, as they used to say back in elementary school, live up to her true potential. For that she needs your unflinching help. Sure, everyone says they want to be the best they can be—tomorrow or the next day, you know, when the time is right. *Hooey.* There's no right time; there's only now. And right now, your job is to see that circumstances beyond your protagonist's control fling her out of her easy chair and into the fray. A story is an escalating dare, and its goal is to make sure your protagonist is worthy of her goal. This means that, as difficult as it may be, when it comes to the care and feeding of your protagonist, you have to be mean to her. Hold her soles to the fire, even when she starts to squirm. Even after she cries, "Uncle!" After all, the last thing you want is a hero who is all hat and no cattle.

But wait, you may be thinking, *that's just true of commercial fiction, isn't it?* Commercial fiction, they say, is plot driven, so lots of stuff has to happen, and it has to build and have consequences. Literary novels don't really need something as contrived and surface as an actual plot, since they're character driven. Slice of life and all that. Right?

Actually, wrong. Very wrong, in fact.

MYTH: Literary Novels Are Character Driven,
So They Don't Need a Plot

––––––

REALITY: A Literary Novel Has Just As Much Plot
As a Mass Market Potboiler, If Not More

Since serious literature is less prone to "big" events than commercial fiction is, it is actually *more* in need of a well-constructed plot than anything Jackie Collins ever dreamed of. In literary fiction the plot must be far more layered, intricate, and finely woven in order to illuminate subtler and more nuanced themes. Character-driven novels rely a lot less on sinking ships, falling meteors, and tidal waves, and a lot more on a missed gesture, a quick nod, a moment's hesitation— which in the hands of a great writer can feel more earth shattering than a nine-point earthquake. But make no mistake: literary fiction still revolves around an escalating series of challenges that the protagonist must brave, because no matter how keenly honed the protagonist, he still has to want something real bad. And if that desire doesn't put him to the test—yes, just as in a potboiler, it's baptism by fire—then he, and the narrative he inhabits, will remain flat and uninvolving. Remember: a story revolves around events that force the protagonist to come to grips with a difficult inner issue—which, ironically, is something literary novels are far more geared to convey. So don't fall prey to this tired old saw; instead, kick it to the curb—poetically, if you must.

Case Study: *Sullivan's Travels*— The Evolution of a Bumpy Night

Okay, we've admitted that yes, no matter how much we love our protagonist, if he wants to be the center of attention in an actual story, he's going to be in for a bumpy night. How bumpy? At first, not very. In the beginning the protagonist's quest tends to look easy—to him, that is. It has to. Because just like in life, if he *knew* the buckets of blood, sweat, and tears his hard-won triumph would require, he probably wouldn't even get out of bed. Luckily neither we, nor our protagonist, ever know how hard it's going to be. Take, for instance, John L. Sullivan, the privileged young film director and protagonist of Preston Sturges's classic 1941 film *Sullivan's Travels*. Tired of directing successful yet meaningless pieces of fluff—and you only have to hear the title of his latest film, *Hey-Hey in the Hayloft,* to get the picture—Sully wants to direct a serious drama. "I want this to be a picture of dignity . . . a true canvas of the suffering of humanity," he says, brushing off his worried producer's hopeful question, *"But with a little sex in it?"*[6]

When it's pointed out to Sully that he has no actual experience in suffering of any kind, he instantly agrees, but instead of giving up, he decides there's a simple solution. He'll suffer. How hard can it be? So he goes to the wardrobe department, picks out sufficiently raggedy clothes (which he dons with his butler's help), then hitchhikes out of town with a dime in his pocket. But rather than suffering, he experiences only mild annoyance at the hands of a middle-aged man-crazy widow and soon finds himself back in Hollywood.

Realizing this suffering business isn't as easy as poor people make it seem, he sets out again. But the studio, now worried that he might actually find the trouble he's so determined to get into, insists that a large mobile home chock full of "babysitters" follows him—just in case. This time the only thing he suffers is fools. When this doesn't work, Sully balks and ups the ante, hitting the road again, at last riding the rails with actual hobos. Now he sees genuine suffering and devastating poverty. He sleeps on the floor; he goes hungry. But there's a big difference between being poor and being broke, especially when back home, you're rich. Strike three. This time his plan doesn't work because he's too uncomfortable to stay uncomfortable long enough to get the hang of it.

Now Sully really is ready to throw in the towel, return to Hollywood, and sort things out. Everything he tried backfired, so what's the use? Besides, he's begun to suspect that there's something sordid about being a voyeur at the table of human suffering. It feels too much like tempting fate. And in the-beware-of-what-you-wish-for category, that's exactly when life steps in and raises the stakes, big time. A hobo steals Sully's shoes—one of which has a studio ID card sewn into the sole—and is pulverized by a railroad train. The cops, finding the ID card, announce that Sully is dead.

However, the actual Sully has been beaten and robbed of the five-dollar bills he'd been giving out to the hobos before returning to Hollywood. In a stupor, he assaults a railroad cop and is arrested. He tells them who he is, fully expecting that to be that. But without ID, and the headlines full of the news of his death, who would believe him? No one. Sully is convicted and sent to a prison work camp where, at last, life bestows upon him the very experience he'd been seeking: human suffering without an escape clause. Goal met. Now, when he gets back to Hollywood, he'll have the know-how to make a picture about genuine human suffering.

Except the lesson he ultimately learns is the exact opposite of what he'd expected. Because now he knows firsthand that the last thing suffering people want to watch is more people suffering. What they want is a *break* from suffering. They want to laugh, and for a moment *forget* about everything that's wrong in their lives. They want to watch movies like *Hey-Hey in the Hayloft* and feel how wonderfully silly life can be.

And so, in the end, because everything that could go wrong, did— and then some—Sully has the experience that a perfect story bestows upon its protagonist: he returns to the place where he began and sees it with new eyes. The world didn't change. He did.

Had writer-director Sturges shown Sully mercy, the film could have ended when Sully realized that, try as he might, there's just no way he'd ever have a clue what it feels like to be disenfranchised. And hey, he did try pretty hard, didn't he? So it would have been a job well done, right? Nope. Because until Sully finds himself in prison with no way out, everything has been on his terms. And a test on your own terms is no test at all. Sturges knew this, so rather than swooping in at the eleventh hour and saving Sully from the chain gang, he stepped back and let life have a whack at him. In so doing, he actually did Sully a huge favor. As the saying goes, "No man is more unhappy than the one who is never in adversity; the greatest affliction of life is never to be afflicted." Only by making sure Sully was *extremely* afflicted did Sturges give him the opportunity to become a better man.

The Importance of Hurting the One You Love

While getting writers to punch, shoot, stab, and otherwise rough up their protagonist can be difficult, there's something even harder to get them to do: embarrass their hero. After all, a punch is a punch; it's physical, external—once the sting fades and the wound heals, it's usually gone and forgotten. What's more, physical pain is something one can keep to oneself. No one else has to know. But to embarrass someone? That's public. Unlike physical pain, embarrassment says something about you—it means that you not only made a mistake, but that you were found out. Social pain—embarrassment, mortification, shame—lingers; the full measure of its sting tends to be felt afresh every time you think about it, even though decades may have passed.[7] It's no surprise the word *mortify* originally meant "to die," because that's often exactly what we want to do when we're embarrassed.

It also tends to be the thing that best spurs growth.

So it's a pity that embarrassment, mortification, and shame are the last thing writers want to put their protagonist through. We don't need to read *Pygmalion* to know writers and artists have a tendency to fall for their creations. So, without meaning to, they're always smoothing the way for him, pitching softballs—sort of like an attentive director always making sure the camera only catches the star's "good side." In real life, it's bad form to put someone in an awkward situation—worse still, to then point the finger at him and make sure everyone notices.

After all, it's one thing to fail in private and quite another to fail on the page in plain view. Like if John graduates from a prestigious law school, then fails the bar. Twice. And he's thinking, *Well, at least no one knows but me.* Except when he's John F. Kennedy, Jr., and it's the headline of the *New York Post*, which actually read: "The Hunk Flunks." Failing in public is mortifying. But it sure triggers change, whether that means adopting an alias and moving to another state where you can pretend you're someone else, or doing as Kennedy did and rising to the challenge. (For the record, he stuck to it, passed the bar, and went on to win all six of his cases as a prosecutor for the Manhattan district attorney's office.)

Constantly upping the ante gets the protagonist in shape, which is crucial, since the final hurdle he'll have to sail over will be impossibly high. Thus the more you put him through before he gets there, the better. After all, as Emily Dickinson points out, "A wounded deer leaps the highest."[8] If you want your protagonist to be up to the test when he gets to that last hurrah, you've got to toughen him up along the way.

Keeping in mind that your reader must know what your protagonist's plan is before you begin to dash it, here's a crash course on how to torture your protagonist—for his own good, naturally.

Eleven Do's and Don'ts for Undermining Your Characters' Best-Laid Plans

1. **Don't let your characters admit anything they aren't forced to, *even* to themselves.** Remember when you were a kid, and someone was trying to get you to do something you didn't want to do? You'd yell, "Oh yeah? Make me!" Well, in a story, when it comes to admitting anything, *ever*, that's your characters' mantra. No one in your story should ever divulge anything they

aren't forced to—either by a gun to the head or, far more likely, circumstances beyond their control. Information is currency. It has to be earned. No one gives it away for free—and everything has a price. Your protagonist needs a compelling reason to admit anything. It either gains him something or keeps something bad from happening. It's never neutral.

2. **Do allow your protagonist to have secrets—but not to keep them.** We keep secrets for one reason: because we are afraid of what will happen—that is, change—if they're divulged. But that doesn't make it easy. A secret is "the result of a struggle between competing parties in the brain. One part of the brain wants to reveal something, and another part does not want to," writes neuroscientist David Eagleman in *Incognito: The Secret Lives of the Brain.*[9] In fact, turns out it's unhealthy to keep a secret, both mentally and physically. According to psychologist James Pennebaker, "the act of *not* discussing or confiding the event with another may be more damaging than having experienced the event per se."[10]

Thus, given how painful it can be to torture your protagonist, it's comforting to know that ultimately forcing her to divulge her secret will actually be a kindness. You don't want her to have a heart attack from the stress of keeping it in, do you? So no matter how fervently she may want to keep her secrets close to the vest, you can't allow it. In fact, the more the protagonist wants to keep mum, the more the story will try to make her sing.

And one more thing: don't keep her secret a secret from us—let the reader in. We love being insiders. Our delight comes from knowing what the protagonist is holding back and why; we revel in the tension between what she's saying and what we know she's really thinking.

3. **Do ensure that everything the protagonist does to remedy the situation only makes it worse.** This is otherwise known as the irony factor. Remember what we said about the decision in one scene triggering the action in the next? This is how it plays out, ever upping the stakes, forcing the protagonist to reevaluate the situation with each turn of the screw.

There are myriad ways to up the ante. For instance, April is secretly in love with Gary, so she applies for a job at his firm to get to know him better. She's hired, and in Gary's department, no less. But when she shows up for work all decked out in a new outfit she can't really afford, she discovers she's actually gotten Gary's job. He, it turns out, has been promoted and is being transferred to the London office. (Or worse, he's been fired, because her experience was so much stronger than his.)

Sometimes the irony stems from the fact that the plan works brilliantly and the protagonist gets *exactly* what she's after, only to discover it's actually the last thing she'd ever want. In which case, Gary instantly falls for April, sweeping her into his arms, murmuring that he loves her almost as much as playing *World of Warcraft* until dawn, which he'd do every night if only his mom would stop banging on the wall.

4. **Do make sure everything that can go wrong does.** But don't let your protagonist in on your agenda. Let him start out believing all he has to do is ask, and voilà! All the riches in the world will be delivered by FedEx before nine the next morning. It's not that he's delusional; it's human nature. As we know, in order to conserve precious energy, anytime the brain can do less, it will,[11] and we follow suit. In the beginning, no one ever spends more than the minimum effort required to solve a problem. But honestly, can you remember the last time the smallest amount of effort solved *anything*? In fact, it's practically guaranteed to make things worse, and hopefully in ways

the protagonist never imagined. That's why we cringe in movies when the hero breathes a sigh of relief and says, "Well, at least nothing *else* can go wrong." Because we know that can mean only one thing: now something *really* bad is going to happen—and usually it's something that makes everything up to that moment seem like a cakewalk.

5. **Do let your characters start out risking a dollar but end up betting the farm.** Another interesting facet of the escalating trouble that follows most protagonists is that although they begin by merely betting a lowly dollar, they tend to cower, whine, and fret more about that single dollar than they do at the end, when betting the entire farm. For instance, in the 1986 John Hughes classic *Ferris Bueller's Day Off*, Ferris's sidekick, Cameron, has never stood up to his father—a man who, according to Cam, loves his vintage Ferrari more than life itself. Which is why he never drives it. But because Cam is a wimp— he can't stand up to anyone—he lets Ferris talk him into cutting school and taking the car out for a spin. Ferris assures Cam that afterward they'll simply run the car in reverse to get rid of the couple of miles they'll put on the odometer. Cam wails and moans but hasn't the gumption to say no.

Naturally, instead of a quick spin, they end up driving around all day, racking up far more mileage than Cam ever dreamed, not to mention putting the car in constant danger of being dinged, lost, or stolen. Cam begins by whining, but as the day progresses, and he finds himself in situations that force him to toughen up, he realizes he has far more grit than he thought—*and* that keeping such a magnificent car enshrined in a glass garage rather than taking your chances driving it is, at best, foolish (as is lavishing more attention on a car you don't drive than on your son). Thus at long last, Cam finally gets mad at his dad.

Even so, Cam is a bit panicked when at the end of the day they discover, not surprisingly, that putting a car up on blocks and wedging the gas pedal down with the transmission in reverse doesn't, in fact, take the mileage off. Furious, Cam finally unleashes his pent-up anger by kicking the front of the car, denting it. Realizing he's now ready to stand up to his dad, with a satisfied smile he leans on the car, accidentally knocking it off the blocks. With the engine racing, the second the tires hit the ground and gain traction the car crashes through the garage's glass wall and sails out, plummeting into the ravine below.

Which brings us to the fabled Aesop, who said, "Men often bear little grievances with less courage than they do large misfortunes." And so, having learned to stand up for himself throughout the day, rather than accepting Ferris's offer to take the blame for the wrecked Ferrari, Cam digs deep and finds the courage to tell his father what happened. He is far less fearful of telling him the truth—with the car in pieces at the bottom of the hill—than he was that morning, when the worst thing he thought he'd have to confess was that they'd put ten miles on the odometer.

6. **Don't forget that there is no such thing as a free lunch— unless, of course, it's poisoned.** This is another way of saying everything must be earned, which means that nothing can come to your protagonist easily—after all, the reader's goal is to experience how he reacts when things go wrong. As Steven Pinker points out, stories can help us "expand the range of options in life by testing, in small increments, how closely one can approach the brink of disaster without falling over it."[12] This means the protagonist has to work for everything he gets, often in ways he didn't anticipate (read: that are much harder than anything he would have signed on for). The only time

things come easily is when they are the opposite of what is actually best for him.

For instance, in *It's a Wonderful Life*, out of the blue the villainous Potter summons George into his office and, in a deceptively soothing voice, offers him the opportunity of a lifetime: a job with an outsized salary—which would be an instant way out of his nickel-and-dime existence. George even considers it for a minute. But being far smarter than that simp Snow White (even the *birds* knew better than to take that apple), he knows a poisonous spider when he sees one. He is well aware that if he takes what Potter is offering him, it will cost him big time.

7. **Do encourage your characters to lie.** While in real life, we don't want people to lie to us, in a story, characters who lie are the ones who catch our interest. A provocative lie can make even the most bland character intriguing because we then think, *Hmmm, I wonder why she lied. What's she got to hide? Maybe she's not so bland after all.*

This, of course, means you need to let us know the character *is* in fact lying. If we don't know it's a lie, how can we anticipate what will happen when the truth is discovered? Because like secrets, lies, once told, must eventually be exposed. In fact, a big part of what keeps the reader turning pages is imagining the lie's possible consequences.

Are there times when a lie doesn't get found out? Of course. But never "just because." Rather, the reason the lie is left unexposed must tell us something important about the characters. And sometimes the fact that the protagonist gets away with something *is* the story. For example, in Patricia Highsmith's brilliant novel *The Talented Mr. Ripley*, the protagonist, an amoral young man named Tom Ripley, is soon a murderer. Since there are five Ripley novels in all, it's not giving anything away to say Tom does not, in fact, get found out—which means he lies all the way through. Thus the thrill of the novel comes from his fear that his lies *will* be exposed, juxtaposed with our anticipation of how and why they *won't*. This is a perfect example of screenwriter Norman Krasna's maxim "Surprise 'em with what they expect."[13]

This brings us to the one person in a story who must not lie, no matter what: you, the writer. Yet writers lie all the time, often because they don't want the reader to "figure it out" yet, as we discussed in chapter 6. The trouble is, the reader has implicit trust in you, so when she discovers you lied to her, she starts wondering what *else* in your story might not be true, and she begins to suspect everything.

8. **Do bring in the threat of a clear, present, and escalating danger—not a vague facsimile thereof.** Everyone knows you need a force of opposition. Without one, the protagonist has nothing to play against, making it damn near impossible for him prove his worth, no matter how hard he tries. Which is why the force of opposition must be well defined—and *present*. It can't be a nebulous threat that never really materializes, or an antagonist, no matter how potentially dastardly, who merely hovers meaningfully on the edge of the action but never actually *does* anything.

To that end, there is one accessory that no antagonist should leave home without: a ticking clock. Nothing focuses the mind—not to mention the actions of the protagonist—better than a rapidly approaching deadline. This not only keeps the protagonist on track, but keeps the writer on track as well, by constantly reminding her that as much as she'd love to send the protagonist off on a soul-searching weekend in Tuscany, unless he finds Uncle Milt's will by midnight, all will be lost when the wrecking crew arrives at dawn.

Of course, the force of opposition doesn't have to be a person. It can be conceptual, like the straitjacket of strict social conformity, the dehumanization of unchecked technology, or the tyranny of the letter of the law. But—and it's a big but—it can't *stay* conceptual because, as we know, concepts are abstract; they don't affect us, either literally or emotionally. What *does* affect us is a concept made specific and thus concrete. This means the concept needs to be *personified* by specific characters who try to force the protagonist to bend to their will.

For instance, Ken Kesey's novel *One Flew Over the Cuckoo's Nest* is about how the demand for social conformity straitjackets those bent on following their own drummer and, if that doesn't work, lobotomizes them. In the story, which takes place in a mental hospital, these things play out literally, spurred by an antagonist aptly named Nurse Ratched. Although she's the one who wreaks havoc on the lives of the men in her care, she is merely the personification of the theme, which she nevertheless embodies with ruthless gusto.

9. **Do make sure your villain has a good side.** We already know that, as counterintuitive as it seems, the villain has to have a good side, however fleeting and minuscule. After all, no one is all bad. Or, if they are, they rarely see themselves that way. The majority of history's bloodthirsty, despicable despots, not to mention elected officials, thought they were doing a good thing, often in the name of God and country. But even more to the point, black-and-white characters—whether all bad or all good—are tedious, not to mention impossible to relate to. In fact, sometimes a totally good character is even more off-putting than a bad guy.

Think about it—that ruggedly handsome guy in the office who does everything right all the time, has a perfect family life, and a desk that's never messy—don't you secretly wonder what's buried in his basement? Not out of envy (probably), but because no one could really be that "perfect." Just as the protagonist needs a flaw, so the antagonist needs a positive trait.

What's more, a character who's 100 percent bad isn't likely to change, which renders him one-note. When it comes to "what you see is what you get," what you tend to get is bored. Whereas a villain with a couple of good qualities just might be redeemable, instilling suspense. Not that your bad guy has to *be* redeemed, mind you, but both he—and the story—are far more intriguing if the possibility is open.

10. **Do expose your characters' flaws, demons, and insecurities.**
Stories are about people who are uncomfortable, and as we know, nothing makes us more uncomfortable than change. Or, as Thomas Carlyle said, "By nature man hates change; seldom will he quit his old home till it has actually fallen around his ears."[14]

This means that a story is often about watching someone's house fall around their ears, beam by beam. After all, premises that begin, "I wonder what would happen if . . ." rarely postulate, "a happy, well-adjusted woman was contentedly married to a wonderful, happy man and had a great career and two equally happy, well-adjusted kids." Why? Besides the fact that "perfection" is not actually possible (and thank god for that), things that are not falling apart are dull (unless, of course, it's *your* house, in which case dull is good).

Thus it's your job to dismantle all the places where your protagonist seeks sanctuary and to actively force him out into the cold. Writers tend to be softies, so when the going gets rough, they give their protagonist the benefit of the doubt. But a hero only becomes a hero by doing something heroic, which translates to rising to the occasion, against all odds, and *confronting one's own inner demons in the process.* It's up to you to keep your protagonist on track by making sure each external twist brings him face to face with something about himself that he'd probably rather not see.

11. **Do expose *your* demons.** There's another, trickier reason writers sometimes shield their protagonists and let them duck the really thorny questions. Rather than protecting the protagonist, sometimes it's the *writer* who's uncomfortable with the issue the protagonist faces. By allowing the protagonist to sidestep it, the writer, too, gets to avoid it. Because just as you "out" your characters, so will they out you. After all, if you make them do things propriety frowns on, you're revealing that you're no stranger to the uncivilized side of life yourself— that is, all those things we do and think when we're pretty sure no one else is looking. This, of course, is precisely what the reader comes for. We all know what polite society looks like—no one needs to explain it to us; we get it. But beneath our very together, confident public persona, most of us are pretty much raging messes. Story tends to be about the raging mess inside, the one we struggle to keep under wraps as we valiantly try to make sense of our world. This is often the arena the *real* story unfolds in, and what causes the reader to marvel in relieved recognition, *Me too! I thought I was the only one!* And so, to both the writer *and* the protagonist, Plutarch offers this sage advice: "It must needs be that those who aim at great deeds should also suffer greatly."[15] Often in public.

Or, to put it a bit more philosophically, there's Jung: "One does not become enlightened by imagining figures of light, but by making the darkness conscious."[16]

CHAPTER 9: CHECKPOINT

Has everything that can go wrong indeed gone wrong? Don't be nice, even a little bit. Throw social conventions out the window. Does your plot continually force your protagonist to rise to the occasion?

Have you exposed your protagonist's deepest secrets and most guarded flaws? No matter how embarrassing or painful the revelation, have you forced him to fess up? Have you obliged him to confront his demons? How can he possibly overcome them (or realize that they aren't so bad after all) unless he's forced to come to grips with them?

Does your protagonist earn everything she gets, and pay for everything she loses? This is another way of saying that there must be a consequence to everything that happens. Ideally, it's a consequence that forces your protagonist to take an action she'd really rather not.

Does everything your protagonist does to make the situation better actually make it worse? Good! The worse things get for your protagonist, the better for your story. By making sure that things go from bad to worse, you will keep your story's pacing on track as the tension—and the stakes—ratchet ever upward.

Is the force of opposition personified, present, and active? It doesn't always have to be a giant, raging gorilla or a gun-toting psychopath, but readers want someone (or something) to root against. This means that vague threats, generalized "evil," or unspecified possible disastrous events don't cut it. The danger needs to be specific—and wired to a rapidly ticking clock.

THE ROAD FROM SETUP TO PAYOFF

COGNITIVE SECRET:

Since the brain abhors randomness, it's always converting raw data into meaningful patterns, the better to anticipate what might happen next.

STORY SECRET:

Readers are always on the lookout for patterns; to your reader, everything is either a setup, a payoff, or the road in between.

Art is the imposing of a pattern on experience.

—ALFRED NORTH WHITEHEAD

RED ABOVE THERE JOKES GRAVEL, *instant might round most.* Hard to read, huh? It feels like a train wreck inside your skull. With each new word it further defies the linguistic pattern you innately expect, which means no extra dopamine for you; instead, your neurotransmitters give you less of it than normal, in an effort to express their—that is, your—displeasure.[1]

Your brain doesn't like anything that appears random, and it will struggle mightily to impose order—whether it's actually there or not. Take a starry, starry night, for instance. As Nobel laureate in physics Edward Purcell wrote to evolutionary biologist Stephen Jay Gould, "What interests me more in the random field of 'stars' is the overpowering impression of 'features' of one sort or another. It is hard to accept the fact that any perceived feature—be it string, clump, constellation, corridor, curved chain, lacuna—is a totally meaningless accident, having as its only cause the avidity for pattern of my eye and brain!"[2]

But one thing that isn't random is our passion for patterns, even if we do get carried away sometimes and see the face of our beloved etched in the clouds. Searching for patterns is a habit that began long before indoor plumbing, refrigerators, and doors, when home consisted of a nice cave, with maybe a comfy pile of leaves to bed down on, and being able to predict what might happen next was often a matter of life and death. Since lions and tigers and bare cavemen—oh my!—could walk in unannounced anytime, day or night, the brain had to become expert at translating any and all data into patterns, allowing us to determine what that bump in the night might be. After all, unless we know what the normal pattern is, how can we possibly detect something out of the ordinary? "The brain is a born cartographer," says neuroscientist Antonio Damasio.[3] From the moment we leave the womb, it begins charting the patterns around us, always with the same agenda: *What's safe, and what had I better keep my eye on?*[4]

Stories are about the things we need to keep an eye on. They often begin the moment a pattern in the protagonist's life stops working—which is good, because, as scholars Chip and Dan Heath note, "The most basic way to get someone's attention is this: Break a pattern."[5] Can you see the fine print? In order to break a pattern, we need to know what the pattern is. And as far as the reader is concerned, *everything* is part of a pattern—and the thrill of reading is recognizing those patterns. What's more, the reader assumes there is an interrelationship among all the facets of a story—that the patterns interlock in the same way ecosystems, borders, and jigsaw puzzles do. Yet this is the level of story that writers sometimes dismiss as mere plot, while toiling away making sure they've woven in a perfectly nuanced leitmotif based on water dripping and a big spatula. It's kind of like piling frosting onto a cake you haven't baked yet. Because, while readers may savor nuance, unless it illuminates and deepens a clear-cut pattern they've been following, it's nothing more than fancy window dressing in a vacant house.

It should be pretty evident by now that readers are a very demanding lot. We have specific expectations (which we're rarely consciously aware of), and our brain wants them met or we're taking our ball and going home. One of our most hardwired expectations is that anything that reads like the beginning of a new pattern—that is, a setup—will, in fact, *be* a setup, with a corresponding payoff. What's more, we have a voracious appetite for setups. We love them because they're intoxicating; they stimulate our imagination, triggering one of our favorite sensations: anticipation. They invite us to figure out what *might* happen next, which leads to an even *better* sensation: the adrenaline-fueled rush of insight that comes from making connections ourselves.[6] When we identify a setup, guess what will happen, and end up being right, we feel smart. Setups seduce us with the granddaddy of all sensations: engagement. They make us feel involved and purposeful, like we're part of something—and an insider to boot. Readers see setups as the writer's way of talking in code. We know from the instant we spot one that it's now our job to feverishly track the pattern leading to the payoff; we tackle it with abandon, relishing every moment, even when it keeps us reading long past our bedtime.

To make sure your stories have lots of tired but satisfied readers, in this chapter we'll explore just what a setup is and how to make sure the road from setup to payoff is actually visible on the page. We'll examine how unintended setups derail a story and take a look at simple setups that pay off big time.

Look Out—I Think It Might Be a Setup!

So what exactly *is* a setup? It is just what the word implies. It's something—a fact, an act, a person, an event—that implies future action. In its most basic form, a setup is a piece of information the reader needs well in advance of the payoff so the payoff will be believable. It can be something as simple as letting us know early on that James speaks Swahili, so when it turns out the instructions for diverting the meteor before it slams into downtown Des Moines are written in Swahili, we won't groan when James announces he can read them. It also means that because readers won't know the *real* reason you're telling them about James' bilingual prowess in chapter one, there needs to be a credible story reason for it to come up at that moment so it's not a total giveaway, in neon, that you're Trying to Tell Us Something. There's a fine line between giving the reader a tantalizing bit of information that piques her imagination and clobbering her over the head with something so obvious there's no suspense whatsoever. Arouse her suspicion, though, and she'll love you for it.

What you want the reader to think is *Gee, I understand James speaks Swahili because it was the only language offered in his high school, and he couldn't graduate without it, but something tells me that by the end of the story he's going to be glad it worked out that way.* Which means, of course, that if the whole Swahili thing *doesn't* come up again, it will turn into one of those lonely elephants, wandering the halls of the story, looking for something to do (damage, most likely).

Setups are, of course, often far more intricate and involving than James's speaking Swahili—which is, after all, just a single piece of supporting information. Often a setup triggers an entire subplot, motive, or way to interpret what's happening, as we'll soon see. That said, it's important to point out from the get-go that the payoff the reader then anticipates doesn't have to be correct. Far from it. Very often the *true* meaning of the setup is clear only in hindsight. As we discussed earlier, in Hitchcock's masterpiece *Vertigo,* we're set up to believe the enigmatic Madeleine is the beautiful, disturbed wife of the man who hired ex-cop Scotty Ferguson to protect her, only to later discover that she is actually a shopgirl who's been hired to pose as Madeleine. As we've already seen, the catch in this type of scenario is that when the payoff comes, everything that happened up to that moment to support the false assumption must now, in hindsight, support the new twist. As Raymond Chandler wisely noted, "The solution, once revealed, must seem to be inevitable."[7]

There's no avoiding this truth. So keep in mind that to the reader, *everything* in a story is either a setup, a payoff, or the road in between.

Setups That Aren't

Readers are always on the lookout for patterns, so the last thing you want is for the reader to decide something is a setup that isn't—and worse, to act on it. It's like when the creepy guy in the next cubicle decides that the way you always ignore him proves you're secretly in love with him, so of course he should make his move. In a story, this translates to dragging some irrelevant bit of information through your otherwise carefully constructed tale, undermining the assumptions that you *do* want the reader to make.

I can't stress this too strongly: readers' cognitive unconscious assumes that everything in a story is there on a need-to-know basis, so they take it for granted that everything you present is part of a pattern. They believe that each event, fact, or action will have critical significance. Thus it's astonishingly easy to mistake a digression or a random unnecessary fact for a setup. To make matters worse, because its relevance to what's happening *now* seems shaky, readers take that to mean it will have even more significance *later*. And so it becomes part of the filter through which they run the meaning of everything that happens from that point on.

For instance, let's say the protagonist, Nora, mentions in passing to her husband, Lou, that Betty from next door spent all day loudly berating that good-for-nothing gun-toting boyfriend of hers. Now, as far as the *writer* is concerned, Nora might have only brought it up to explain why she has such a pounding headache and so can't help Lou search for their missing Labradoodle pup, Rufus. But chances are the *reader* will think, *What? Betty has a gun-toting boyfriend? I bet he has something to do with that poor pup's disappearance. And hey, what happened to Nora's sister, Kathy? She hasn't been around for a while, not since that night she had dinner at Betty's; I wonder if. . . .*

Or worse, the reader can't even figure out how the information *might* apply. What would a gun-toting thug be doing in a gated Quaker commune, anyway? Thus part of the reader's mind lags behind, busily chewing over what the gun-toting guy could possibly mean, while another part soldiers on, reading about Nora and Lou's puppy. But as researchers at Stanford have proven, contrary to popular wisdom, effective mental multitasking is not actually possible—the brain, as it turns out, can't process two strings of incoming information at the same time. According to neuroscientist Anthony Wagner, when trying to focus on multiple sources of information coming from the external world or emerging out of memory, people are "not able to filter out what's not relevant to their current goal."[8] Thus, while the reader's mind mulls over the import of Betty's boyfriend, the significance of what's *actually* happening on the page begins to fade. It's a bit like listening to someone speaking in a very heavy accent. You have to strain so hard to make out the words, you miss what they're trying to tell you. Soon the reader has no idea what's going on, and not long after, ceases to care.

It's not even a choice; it's innate: the brain is wired to go offline—that is, ignore the real world and slip into a fictional one—only if it believes the story will be of benefit by providing info that'll help it navigate this cockeyed world of ours. Once engaged, it flips the switch that filters out actual reality. When that belief is shattered—say, by setups that go nowhere—reality floods back in.[9]

With that in mind, the question becomes: what exactly does a real setup look like? Let's pal around with a few of them to get the feel of what we're talking about.

Case Study: *Die Hard* and *Girls in Trouble*

Sometimes setups don't read like setups at all. For instance, *Die Hard* opens with protagonist John McClane on a plane that's just landed at LAX. A New York City cop, McClane refused to relocate when his wife got a big promotion that required that she move to LA. She took the kids and left. He hopes to win her back. He's exhausted—and clearly glad that the plane is no longer airborne. His seatmate, an older salesman, notices McClane's relief and pegs him as a novice flyer. He then gives McClane a sage bit of advice on how to beat jet lag: stand on a rug barefoot and "make fists with your toes."[10] *Uh-huh*. It's a nice bit of comic relief, and McClane's polite but skeptical reaction tells us something about how he sees the world.

The scene's subtext is clear: beneath McClane's dislike of flying is an even deeper dislike of being out of his element. New York City is a far cry from sunny Los Angeles, especially on the day before Christmas. The question is, is there *anything* about his interchange with the salesman that screams "Setup!"? Not really. It's told us something about McClane, and technically that's enough. What's more, there's no red flag saying, *Look at me—I matter more than you think,* which is great, because that's the last thing you want a setup to do. And this *is* a setup.

Because once McClane gets to the Christmas party at his wife's office and finds himself alone and tense in a plush executive bathroom, he takes his shoes off and tries it. He smiles, amazed that it really works. He's still blissfully making fists with his toes when he hears gunfire. With no time to do anything but grab his Beretta, he rushes into the hallway to check it out. *Barefoot.*

And so he spends the rest of the movie running through minefields of broken glass, bleeding. That opening scene? Beyond being entertaining, and telling us something about McClane, it was a setup that paid off by making his road to becoming a hero that much more difficult.

You might wonder whether setups like that are worth the trouble. Couldn't McClane simply have taken his shoes off anyway? Couldn't he have muttered something about how your feet seem to swell while you fly, and it's such a relief to slip off your shoes for a minute? Absolutely. He even could have accidentally splashed water on them as he washed his face, and so taken them off to let them dry. But for the audience, both of those scenarios would have lacked something that the opening scene provided: a small "aha!" moment—that delicious sense of understanding when we grasp the specific (and sometimes deeper) *why* behind a character's action. It's this that allowed us to savor the irony: *If only that guy on the plane had kept his trap shut, McClane wouldn't be leaving such a bloody trail everywhere he goes.*

Setups, when done well, read like fate.

While this is a minor twist in *Die Hard*, Caroline Leavitt's novel *Girls in Trouble* offers a far more fleeting, yet substantial example. The novel begins in Boston and revolves around the relationship between Sara, an unmarried sixteen-year-old who is pregnant, and George and Eva, the couple who adopt her baby. It'll be an open adoption, they promise. Sara will be welcome anytime. And before the birth, this is indeed true. George and Eva, hungry to spend as much time with Sara as possible—both because they genuinely care for her and because they fear she might change her mind—shower her with love and attention. However, once the baby is born, Sara's dependence on them becomes overwhelming. What's more, her presence begins to threaten Eva, who wants to feel that she alone is the baby's mother. There is a palpable sense of trouble brewing, and the reader is sure that sooner or later it will come to a head.

In the midst of this, George, a dentist, feeling the strain of his new situation—his unexpectedly fierce love for the baby, Sara's neediness, Eva's growing need to distance herself from Sara—reflects on his day:

> At four, he was finished, an hour earlier than he had hoped. His last patient had been another emergency, a woman who had come in with her bridge still attached to a bright red taffy apple she hadn't been able to resist biting. She left with a temporary and a list of the foods she shouldn't eat. He'd have to place an ad for another hygienist. He wished he could place an ad for a clone. Most dentists worked solo, and he had never wanted to be in a partnership, but maybe it might help things. He wouldn't have to work so hard, such long hours. But of course the question was, who would be the partner? You had to be careful with things like that. The only person he could think of was his old friend Tom from dental school, who lived in Florida and was always trying to get him to move down there. "Blue skies, sandy beaches," Tom urged, but George hadn't really wanted to move.[11]

This passage is on page 98. George doesn't think of Tom, or Florida, again until page 169. But throughout the seventy pages in between, the reader does. Because that one offhand reference to Tom's urging George to move to Florida leaps out as a setup, a revelatory detail that puts the reader on notice. We sense this, even though there is absolutely nothing in the way it's presented that says, *Pay attention, remember me.* But we do anyway. Why? Because the story itself has already supplied us with a context—a pattern—into which this tidbit fits neatly. We know how untenable the situation is, and that it's only going to get worse. We've been anticipating that something will have to give, but until George thought about Tom in Florida, we weren't sure how it might actually play out.

From that moment on, we suspect that George and Eva will move away. And we begin anticipating how Sara will react when they do. Thus this small setup, this tiny nugget, takes on a much greater significance, affecting how the reader interprets everything that happens in the seventy pages between the setup (George's first fleeting thought of Florida) and the payoff (when he once again remembers Tom, buys Tom's dental practice, and moves his family to Boca Raton without a word to Sara).

This isn't to say that there aren't times that the thought of Florida might slip the reader's mind or when she might wonder whether George and Eva are really going to move after all. If setups always made specific payoffs inevitable from the moment they appeared, they'd stifle anticipation rather than spur it. What they often do is illuminate a possibility. Sure, that possibility might be exactly what happens—George, Eva, and the baby do move to Florida—but along the road from setup to payoff, the reader always has the sense that it might go either way. What keeps us reading is the building desire to find out.

The Importance of the Highway between Setup and Payoff: Three Rules of the Road

We know anticipation feels really good, and that what readers love to hunt for is the emerging path from setups to payoffs. After all, a big part of the pleasure of reading is recognizing, interpreting, and then connecting the dots so the pattern emerges. To make that possible, there are three basic rules it behooves writers to know.

RULE ONE: THERE MUST ACTUALLY *BE* A ROAD

This means the setup is not allowed to piggyback on the payoff. Piggybacking occurs when we learn about a problem *at the moment* it's been solved. Talk about draining the tension, killing the conflict, deflating the suspense, and making sure the reader has nothing to anticipate! Thus, we hear that Amy's front tooth has been successfully reattached at the very moment we learn both that Morris accidentally knocked it out last night during a rousing game of gin rummy, and that if the dentist hadn't been able to schedule emergency surgery, Amy's lifelong dream would have been dashed, because she would've had to try out for the Miss Perfect Smile Contest this morning sans her front tooth. Great for Amy, boring for us.

Now imagine the tension, conflict, and suspense had we been there the instant the tooth flew out of Amy's mouth, knowing it's only six short hours until the Miss Perfect Smile Contest, wondering how she'll ever find a dentist at one a.m. in Peoria, not to mention what this will do to her relationship with Morris, which was pretty shaky to begin with.

RULE TWO: THE READER MUST BE ABLE TO SEE THE ROAD UNFOLD

This means it can't take place off the page, shrouded in secrecy. There are three reasons writers tend to keep the road between setup and pay-off veiled, if not totally obscured. One, as we already know, is because they're saving it all up for the big reveal. Second, they simply don't realize they're doing it. Thus they set up a promising storyline and then leave it to the reader to imagine how, specifically, it plays out, until somewhere down the line it resurfaces just in time for the big payoff. Often these writers are under the mistaken assumption that by letting readers know what's going on, they're somehow talking down to them, and so the bulk of the story remains in the writer's imagination.

Thus we learn that John needs to marry before his thirtieth birthday in order to get the inheritance he's been counting on. Then, over the next several hundred pages, John goes on dates whose specifics we never hear—dates we wouldn't be able to interpret anyway, because we have no idea what he's looking for in a wife, or even whether he wants to marry at all. Then, at some point, John decides to marry someone for some reason, and he gets a whole lot of money. The End. Except chances are the reader will never know it, because it's highly unlikely she'll have stuck around that long. Point being, while we're eager to connect the dots, we don't want to have to invent them first.

And this brings us to the third reason writers sometimes inadvertently skimp on the "tells" necessary to establish a pattern. As the author, you know everything about your story—where it's going, who's really doing what to whom, and where the proverbial (and sometimes literal) bodies are buried. Because of this, you're acutely aware of exactly what each "dot" really means and how it all fits together. But here's the thing: your reader isn't aware. What comes across to you as so utterly obvious that it will "give the whole thing away" is a tantalizing "tell" to the reader, who's counting on such "tells" to be able to do what readers love best: figure out what is really going on.

RULE THREE: THE INTENDED PAYOFF MUST NOT BE PATENTLY IMPOSSIBLE

I don't mean impossible in the "he'll try it and when he fails, it will teach him something" sense. I mean, literally impossible, so that if the protagonist himself had given it a moment's thought, he'd have realized how ridiculous such an endeavor would be.

So how did the writer miss it?

Because the writer knew that something was going to happen to prevent the protagonist from taking more than a step or two down that particular path, so she didn't bother to think it all the way through. Why should she?

Because the reader will.

After all, the reader doesn't know the protagonist isn't going to trudge down that path to the bitter end. And if there's one thing we know about readers, it's that they love to anticipate what will happen next. But it doesn't stop there; once they spot a pattern, they test its validity against their own knowledge. Thus they're often way ahead of the protagonist. And when they figure out what the writer didn't—that a particular payoff is not logically possible—they're the ones who may bail.

For instance, let's say that since kindergarten, Norbert has been secretly in love with Betsy, who's completely blind. Unfortunately, Betsy has never thought of Norbert as anything but a good friend. Now, she's away at Harvard, rooming with several hometown girlfriends, all of whom know Norbert and can see. Lonely, Norbert hatches a plan: he'll apply to Harvard, get in, fake a British accent, and woo Betsy as if he were a complete stranger. The writer, however, already knows that Norbert will never get that far, because she's seen to it that Harvard will reject him. Thus it never occurs to her that Betsy's roommates would have instantly recognized Norbert, giving his identity away before he could utter a single *pip pip cheerio*. In short, you must make sure that what your characters *intend* to do is plausible, even if you already know that something unforeseen will thwart them before they can actually do it.

Case Study: *Die Hard*

Die Hard is a perfect story (the blood, gore, and impossible physical feats of derring-do notwithstanding). Why? Because every setup builds to a satisfying payoff. We saw how the barefoot setup-payoff worked, but there are more. Many more. In fact, every main character arcs, every subplot has a resolution. Nothing is wasted, everything is set up in advance, yet there are a million surprises along the way.

Let's look at one particular case in point, to see how setups spur character arcs, motivation, and subplots: Al Powell, the off-duty cop who first responds to McClane's report of gunfire at Nakatomi Plaza, has been a desk jockey since he accidentally shot a kid he'd thought was armed. Powell hasn't been able to draw his gun in the eight years since. When he admits this to McClane, we have the feeling it's something he hasn't told many people. But he offers it up because the two men have developed a bond, and at the moment, things aren't looking good for McClane, survival-wise. So when he asks Powell what got him off the street and behind a desk, Powell tells him the truth rather than hedging.

His admission is a setup. It defines his character arc, by telling us both what his fear is (he's afraid to draw his gun for fear he'll hurt an innocent) and what his desire is (to get back to real police work instead of pushing papers). This is what underscores and drives his subplot, which in turn plays out in service of the main story question: Will McClane, in the course of saving the day, gain enough insight to win back his wife?

Back to Powell. Throughout the film, Powell stands up for McClane against the knuckleheaded powers that be and gives him encouragement when it really does seem that all is lost. And so at the end, when it looks like all the bad guys are dead, McClane comes out of the building and the first thing he does is find Powell, hug him, and swear he wouldn't have survived without him. Powell humbly disagrees. And he means it. He did what he was supposed to do, nothing more. Nothing heroic.

And then it happens. Karl, the one bad guy not quite accounted for, comes blazing out of the building, machine gun in hand. Karl locks eyes with McClane and levels his gun. This time McClane's a dead duck. Except when the resounding *bang!* echoes through the stunned crowd, it's Karl who falls dead. A reverse angle reveals that this time Powell has, indeed, saved McClane. Now, here's the interesting thing. In the script, Powell then says out loud what we're all thinking. That McClane was right—he wouldn't have survived without him, after all.

Except—script be damned—in the movie Powell *doesn't* say that. He doesn't say anything. His eyes, his expression, register something far deeper, something that doesn't have anything to do with McClane at all. They say that, at last, Powell is back in the game. We don't need to be told; we empathize so keenly that we feel it in our bones.

What makes it such a satisfying payoff is that it's so well earned—each dot along the way upped the ante for Powell. But until that moment, although Powell had indeed gone the extra mile for McClane, he hadn't yet been put to the test. By the time Karl comes thundering into the doorway, we know how much McClane means to Powell, and we know what Powell must overcome to protect him. And protect him he does, in one of the most touching moments of deeply felt macho bonding in the pre-bromance era.

CHAPTER 10: CHECKPOINT

Are there any inadvertent setups lurking in your story? Are you sure nothing whispers, implies, or suggests "setup" without actually meaning it? Remember: a great tool for ferreting out unintended setups is our old friend, the "And *so*?" test.

Is there a clear series of events—a pattern—that begins with the setup and culminates in the payoff? Are you absolutely sure none of your payoffs is piggybacking onto its setup? Equally important, are you sure there is an actual pattern of "dots" or "tells" leading from each setup to its payoff?

Do the "dots" build? If you connect the dots between the setup and the payoff, do they add up? Does a pattern emerge? Will your reader see the escalating progression and be able to draw conclusions from it and so, anticipate what might happen next?

Is the payoff of each of your setups logistically possible? Be sure to think each setup all the way through to its logical conclusion, even those (*especially* those) you know your protagonist won't take more than a step or two toward before circumstances (courtesy of you) force him to abandon it.

11

MEANWHILE, BACK AT THE RANCH

COGNITIVE SECRET:
*The brain summons past memories to evaluate
what's happening in the moment in order to make sense of it.*

STORY SECRET:
*Foreshadowing, flashbacks, and subplots must instantly give readers
insight into what's happening in the main storyline, even if
the meaning shifts as the story unfolds.*

Unless we remember, we cannot understand.

—E. M. FORSTER

MEMORY EVOLVED FOR A VERY GOOD REASON: so you can find your car keys when, for the millionth time, they don't seem to be exactly where you *know* you left them. Memory mines the information your brain has acquired in the past for anything that might help it solve the problem you're facing in the present. So it instantly recalls the day the keys slid behind the couch cushion (damn, not there now), the time you left them dangling from the front doorknob (not there now, either), the way your teenage son tends to "borrow" your car while you're sleeping (aha!). Thus, as with everything the brain is wired for, storing information is adaptive: it supports future decisions and judgments that cannot be known with certainty in advance.[1] A category that sums up, well, pretty much everything except death and taxes.

In his book *Self Comes to Mind,* neuroscientist Antonio Damasio speculates that it's thanks to the intersection of the self and memory that consciousness is able to bestow on us its ultimate gift: "the ability to navigate the future in the seas of our imagination, guiding the self craft into a safe and productive harbor."[2] We use the past as a yardstick against which we size up the present in order to make it to tomorrow. What's more, when we do this, sometimes it's our evaluation of the *past* that changes in light of what we've since learned.[3] Memories are continually revised, along with the meaning we derive from them, so that in the future they'll be of even more use.

In other words, memories aren't just for reminiscing. They never were. Memories are for navigating the now. And not just personal memories. Recall what we've said about stories: they are the brain's virtual

reality, allowing us to benefit from the experience of hard-pressed pro-
tagonists.⁴ By the same token, we learn from watching and discussing how others—whether friends, family, or foe—struggled with the
banana peels that life blithely tossed in their path. We get a kick out of
this because it reveals what might happen if we took a similar course of
action without having to actually suffer the pratfall. As Steven Pinker
points out, "Gossip is a favorite pastime in all human societies because
knowledge is power."⁵ Sometimes this knowledge gives us power over
others, and sometimes it gives us the power to make the right decision
when our time comes.

What this boils down to is that the memory of everything we've
done, seen, and read affects, and is affected by, what we're about to do
right now. To quote Tony Soprano's (rather colorful) lament in HBO's
seminal series *The Sopranos*—when his consigliere Sil presses him to
whack his beloved but weak-willed cousin, Tony B.—"All due respect,
you got no f*ckin' idea what it's like to be number one. Every decision
you make affects every facet of every other f*ckin' thing. It's too much
to deal with almost."⁶

This is true of life; this is true of story. And just like Tony, the
writer is required to deal with it, no matter how overwhelming it feels.
The question is, given that all these memories and decisions are influ-
encing your protagonist as she struggles with her issue, how do you,
as a writer, weave it all together? How do you make manifest relevant
bits from the protagonist's past, the events she witnesses that sway
her outlook, and the effect outside forces have on her, whether she's
aware of them or not? What's more, how do you make it seamless and
elegant—that is, without calling a big fat time-out to fill us in?

This is where flashbacks, subplots, and foreshadowing come in.
And *how* they come in—both literally and figuratively—is exactly
what we'll be exploring in this chapter. We'll learn how to weigh new
information against the story question to make sure it's relevant;
examine the three main ways that subplots add critical depth; explore
the role of pacing and timing when it comes to flashbacks, subplots,

and foreshadowing; and discuss how a little judicious foreshadowing can swoop in to save your story from becoming a groan fest.

The Crow's Secret

Here's a delicious paradox: a story is the shortest distance between two points—the point where the story question shifts into play and the point where it's resolved. However, the shortest distance between these two points is often a very circuitous route indeed. That is to say, the crow flies in spirals. Because it's not just about getting from point A to point Z; it's about being aware of everything—past, present, and future, internal and external—that affects the protagonist's struggle, each step of the way.

How *do* you capture the multilayered experience of life on something as two-dimensional as a sheet of paper, in a medium as linear as words? The same way a painting does. By tricking us. Ironically, the only way to evoke the fullness of reality is by first zeroing in on the heart of the particular story you're telling and parsing away all the real-life distractions that don't affect it. The goal is to then weave in relevant elements of the past, ongoing auxiliary storylines, and hints of the future—whether via a subplot, flashback, or bit of foreshadowing—so the reader sees them for what they are (necessary information) rather than what they aren't (dreaded and deadly digressions).

This can be tricky, since timing is everything. Give us an otherwise crucial piece of information too soon, and you neutralize it; it becomes a digression in spite of itself. Give it too late, and it's a groaner. That's why every subplot, every flashback, must in some way affect the story question—that is, the protagonist's quest and the inner struggle it incites for her—*in a way the reader can see in the moment*. Because just as your protagonist always views the present through the filter of the past, so will readers view every subplot, flashback, and bit of foreshadowing in light of the story you're telling.

Subplots: How the Plot Thickens

A story without subplots tends to be one-dimensional, reading more like a blueprint of your protagonist's day than a revealing rendition of it. Subplots give stories depth, meaning, and resonance in myriad ways. They can give the protagonist a glimpse of how a particular course of action she's considering might play out; they can complicate the main storyline; they can provide the "why" behind the protagonist's actions. And in doing so, they can also neatly plug up any otherwise gaping plot holes, introduce characters who will soon play a pivotal role, and show us things that are happening concurrently. Subplots also help set the pace by giving readers a bit of necessary breathing room, allowing their cognitive subconscious to mull over just where the main storyline might be heading.[7]

QUICK, QUICK, SLOOOOOWWWW— HOW TO MEASURE THE PACE

Stated simply and eloquently by literary blogger extraordinaire Nathan Bransford, pacing is the length of time between moments of conflict.[8] While conflict is what drives a story forward, it's often the mounting anticipation of it that has readers so engrossed they forget to breathe. Too much sustained conflict is like trying to live on a diet of nothing but ice cream sundaes. You'd get sick of them in a surprisingly short time (trust me on this), probably right before all that fat and sugar lulled you into a nice long nap. This is the flip side of what we were talking about in the last chapter. Once a pattern becomes utterly predictable, familiar, normal, our attention inherently wanders; it's a biological universal.[9] Readers can only take in so much continuous conflict before they switch from finding it riveting to wondering what's on TV.

The more you stick to a constant heart-pounding tempo, the quicker the story loses its oomph. Look at it this way: Imagine it's ninety degrees. That's hot, right? Now imagine it's been ninety degrees for your entire life, inside and out, everywhere. In that case, ninety degrees wouldn't be hot, it would be normal. And normal, no matter how sweaty, is dull. I remember watching the second installment of the Indiana Jones franchise at the drive-in. Toward the end, when the story disintegrated into one long, monotonous multimillion-dollar chase scene, I was so bored that in order to stop losing brain cells, I spent a very satisfying half hour cleaning out my car. The most exciting moment of the evening wasn't when Indy triumphed over the bad guys (they were already at work on the sequel, so no suspense there); it was finding my favorite pair of sunglasses buried deep in the glove box.

The goal is to set the pace so each burst of intense conflict in the main storyline—each sudden sprint, each unexpected twist—is fueled by the information and insight that's been building since the previous twist. Each time the conflict peaks, you want to back off a bit to give the reader time to take it in, process it, and speculate on its implications, which is often where subplots come in.

SUBPLOTS: THE READER'S EXPECTATION

Subplots invite the reader to leave the recent conflict behind for a moment and venture down a side road that, he believes, will lead back to the story in the near future. The reader is willing to take this jaunt because he trusts that when the subplot returns to the main storyline, he'll have more insight with which to interpret what's happening.

This is the implicit bargain that readers and writers make when it comes to subplots. Readers accept that sometimes the specific story reason for a subplot isn't completely clear at the outset. But they have the tacit expectation it soon will be, which they trust the writer to fulfill. And so they eagerly begin trying to figure out what the subplot has to do with the story question and what its impact will be. You can see where this is headed. It better actually have an impact! I can't say this too strongly: all subplots must eventually merge into—and affect—the main storyline, either literally or metaphorically, or else the reader is going to be mighty disappointed. And while in the olden days, disgruntled readers suffered pretty much in silence, now there's Amazon. The last thing you want are myriad scathing reviews that potential readers "find helpful."

SUBPLOTS: LAYERING THE STORY

We know everything in a story must affect the protagonist in his or her quest, as in: Neil's goal is to get into Yale, so when he fails his senior history class, his heart sinks. The effect is clear, concise, and direct. And that's good. But it could be better. Since nothing spurs readers' mounting interest more than anticipation, giving us the same information—that Neil will fail history—via a subplot not only adds suspense (as we wonder how Neil will react when he finds out), but an intriguing layer of story as well.

For instance, suppose Neil deserves an A in history, but while he's toiling away on his term paper, we hop into a subplot in which his history teacher, Mr. Cupkak, a humorless hardliner, decides to fail the entire class because he's just discovered that an anonymous student posted a video on YouTube photoshopping his face onto a very naked mole rat. We wouldn't see the effect this has on Neil in that particular scene, but because we know what Neil wants—to go to Yale—we'd instantly grasp the effect it will have on him when he finds out. And so when we return to the main storyline—where Neil is just finishing up his term paper, feeling great because he's sure it's the best thing he's ever written, confident it'll get him the highest grade in the class and maybe even land him the coveted valedictorian slot—*we*, on the other hand, are filled with a creeping dread, knowing he couldn't be more wrong; we are bristling at the unfairness of it—and rooting for him to find a way to take the teacher down. We've become his advocate. We're in his corner; we feel protective of him. And, truth be told, we feel just a little bit jazzed that we're in a superior position—after all, we know something he doesn't. We're engaged to the max, complete with a vested interest in what happens next.

However, it helps to keep in mind that although a subplot gets its primary meaning and resonance based on how it affects the main storyline, it has a life of its own. Subplots arc; they even have their own story question that must be resolved. For instance, don't you wonder whether that horrid Mr. Cupkak will get away with failing the entire history class—not to mention how he got to be such a sourpuss in the first place?

But not all subplots directly affect the protagonist. Sometimes their purpose is to give the protagonist necessary insight, the same way a story gives the reader insight: by letting him benefit from the experience of some other poor dog.

MIRRORING SUBPLOTS—AS IN MIRROR OPPOSITES, THAT IS

As we noted in chapter 5, mirroring subplots don't literally mirror the main storyline, because no reader wants to spend time in the department of redundancy department. Rather, they revolve around secondary characters in a situation similar to the one the protagonist finds himself in, and what happens in them doesn't necessarily have a direct external affect on the protagonist. Instead, the effect is internal, in that it changes the way the protagonist sees the situation—because mirroring subplots reveal alternate ways in which the story question could be resolved. Thus they either serve as a cautionary tale or a validation or provide a fresh perspective.

For instance: Let's say the story question is, will Danielle and Perry revive their failing marriage? In a mirroring subplot, their unhappily married neighbors, Ethan and Fiona, might simply throw in the towel and break up. This spurs our protagonists to reconsider their options, and because Ethan and Fiona seem much happier now that they're finally free, Danielle and Perry begin secretly exploring life on their own.

Ah, but as mirroring subplots unfold, they tend to arc in the opposite direction from the main storyline. They often whisper: *This is what you're wishing for; are you sure it's what you really want?* Thus in the end Fiona and Ethan bitterly regret their breakup, triggering Perry and Danielle's realization that maybe sticking with the devil you know isn't such a bad thing after all. Plus, devils can be sort of cute, in the right light.

But whether mirroring or not, all subplots must earn their keep by giving us information we need to know, be it factual, psychological, or logistical, in order for the main storyline to make sense. Here are three ways a subplot can do its job:

1. **Supply information that affects what's happening in the main storyline.** For example, a subplot that establishes that Mr. Cupkak is so reviled he'd be parodied on YouTube, and so mean-spirited he'd fail everyone in his class as a result, will have a direct impact on Neil's quest.

2. **Make the protagonist's quest that much harder.** By failing everyone in his history class, Mr. Cupkak has indeed made Neil's quest infinitely more difficult.

3. **Tell us something that deepens our understanding of the protagonist.** Forget Mr. Cupkak for a minute; what about a subplot in which Neil's grandfather teaches him to clip schnauzers, revealing Neil's innate love of dog grooming—a major not yet offered at Yale? That might make the reader think: *Gee, I wonder if Neil really wants to go to Yale after all?* And so, in the end, the fact that he's going to fail history could turn out to be a good thing.

Cue the Subplot, Key the Flashback

Tell a friend to ask you, "What's the secret of comedy?" and when she gets to the word "secret," blurt out *"Timing!"* Sure makes the point, doesn't it? Truth is, timing is the secret of just about everything, especially subplots and their close kin, flashbacks. The question is this: once you've vetted their viability, how do you know when, exactly, to slide them in and out of the main storyline without accidentally transforming them into digressions? We already know that subplots (ditto flashbacks) sometimes give the reader a breather from the main storyline, often following a strong scene such as a major turning point, sudden revelation, or surprising twist. What we don't know is how to gauge exactly whether the information in the subplot or flashback is relevant at that moment. So, since flashbacks can be entire subplots, let's explore them as we discuss the art of timing.

Flashback—What's the Cause and What Effect Does It Have?

Recently a student told me a writing instructor had made it very clear to him that one of the first rules of writing is never, ever use flashbacks. It reminded me of the time back in elementary school when our teacher told us the best way to stay healthy was to eat lots of red meat, preferably with potatoes. Oh, wait—does that count as a flashback?

Actually, there's a grain of truth in the advice my student was given—advice I suspect was spurred by frustration. I told him I was sure the instructor had simply read one too many stories in which the narrative stops cold for no apparent reason so the writer can step forward and tell the reader something really important that, if we're lucky, we'll need to know later, and if not, that the writer thought was interesting and so threw in for the same reason a dog licks his you-know-whats: because he can. What's worse, those flashbacks were probably full of pure exposition (telling rather than showing) that went on for page after page.

I told him that what the instructor probably meant was, never use a flashback *poorly*. And because that's what most aspiring writers do, she probably figured she had her bases covered. Because poorly done, flashbacks completely derail a story.

There's a footnote. I'd just told that story as a guest lecturer in a colleague's class when, with a throaty laugh, she said, "Uh, that instructor was me." Talk about an adrenaline spike! Lucky for me, she quickly added, "And yep, that's exactly what I meant." She went on to lament the irreparable harm an ill-advised flashback can do to an otherwise engaging story. She's absolutely right.

A poorly timed flashback is like some guy incessantly tapping your shoulder in a movie theater just after the protagonist has lost everything. You have no desire to look away from the screen, knowing that the second you do, the spell will be broken. That's why the guy better be telling you something you need to know right that very minute— like, the theater is on fire, or you've just inherited a million dollars.

The trouble with flashbacks and subplots is they yank us out of the story we're reading and shove us into something we're not quite sure of. It reminds me of Laurie's speech to Steve at the end of *American Graffiti*. He's about to leave for college, and she doesn't want him to go. "You know," she says, "it doesn't make sense to leave home to look for a home, to give up a life to find a new life, to say goodbye to friends you love just to find new friends."[10] Indeed.

That's precisely what a misplaced flashback feels like. Saying goodbye to a story you love just to find a new one. Which, let's face it, you may not love as much, if at all. This is exactly what happens when, knowing that at *some* point we're going to need to know that Pam (mother of Samantha, the protagonist) was raised by wolves, the writer decides that now is as good a time as any to plunk in a flashback of six-year-old Pam stalking prey with the pack. So he randomly inserts it between a scene in which Samantha finally decides to run for mayor and the one where she gives her very first campaign speech. The reader, who was really wrapped up in Samantha's decision to throw her hat into the ring, is initially confused. *Who's Pam, and why is she on all fours with a bunch of wolves?*

At first we try to find the link between the two stories. Is Samantha going to run on an environmentalist platform? Is this a dream maybe?

But the further we venture into the woods with Pam, the more we realize we have a choice to make. We can either forget about Samantha and throw our allegiance behind this new story or leaf through the book until we find Samantha again, skipping over all this wolf nonsense. It feels like we're standing on a frozen lake, and the ice beneath us has cracked neatly in two and is beginning to drift apart. We know we can only straddle both sides for so long before we have to leap onto one or the other—or fall into the water and freeze to death. And since, ironically, the flashback is the one moving forward, that's usually the ice floe we take refuge on.

So off we go with the wolves. And chances are it will get pretty good. Who needs Samantha, anyway? Running with the wolves is much more fun than listening to a newbie's rambling political discourse. But just when our allegiance shifts to Pam, the author deposits us in some stuffy high school auditorium where Samantha is nervously taking the stage. Except now it's Pam we miss. Not to mention that we're still trying to figure out what the foray into the woods has to do with anything, anyway. It doesn't matter if *later on* we find out that Pam is Samantha's mom, because right now, at this moment, we're lost—which means that we very well might not get to later.

But do we really need a long flashback to tell us this? Couldn't a few well-placed snippets of backstory do the trick? Very possibly. And hey, what's the difference between backstory and flashbacks anyway?

Flashbacks and Backstory: One and the Same?

This is a question that often comes up, and the answer is yes, they are. Same material, different uses. Backstory is just that—everything that happened before the story began—and as such it is the raw material from which all flashbacks are drawn. So what's the difference between a flashback and weaving in backstory? It's simple. A flashback, being an actual scene complete with dialogue and action, stops the main storyline; weaving in backstory doesn't. Backstory is, in fact, part of the present.

Neatly woven in, backstory is a mere snippet, a fragment of memory, or even an attitude born of something that happened in the past and runs through the protagonist's mind as he experiences, and evaluates, what is happening to him in the present.

Here is a perfect example from Walter Mosley's novel *Fear Itself*. The novel takes place in Watts in the 1950s, and in this snippet, protagonist Paris Minton is thinking about why he seems willing to continually put his life on the line for his friend, Fearless Jones:

Anyone who knew me and didn't know Fearless would have been surprised that I would have put myself in such a potentially dangerous situation. To the world in general I was a law-abiding worrywart. I shied away from drugs and crap games, stolen merchandise and any scheme that might in any way be construed as unlawful. I never bragged (except about my sexual endowment), and the only time I ever acted tough was to shout at caged animals.

But when it came to Fearless I was often forced to become somebody else. For a long time I thought it was because he had once saved my life in a dark alley in San Francisco. And that certainty did have a big effect on my feelings toward him. But in recent months I have come to realize that something about Fearless compelled me to be different. Partly it was because I felt a deep certainty that no harm could come to me when I was in his presence. I mean, Theodore Timmerman should have killed me on that street, but Fearless stopped him even though it was impossible. But it was more than just a feeling of security. Fearless actually had the ability to make me feel as if I were more of a man when I was in his company. My mind didn't change, and in my heart I was still a coward, but even though I was quaking I stood my ground more times than not when Fearless called on me.[11]

Rendered in stark simplicity that makes it all the more compelling, the passage illuminates Paris's point of view, giving us insight into how he sees the world, what he values, and most important, why. And in so doing, it also gives us several salient details about his past, without once stopping the story.

A flashback does the same thing but presses the story's pause button to do it, demanding the reader's full and complete attention. Which means it better have a clear reason for doing so at that very moment, lest it become another of those poorly placed, ill-thought-out flashbacks that drove my colleague to the conclusion that it's better to ban them altogether than put them in the hands of those who would use them only to shoot their stories in the foot.

Flashbacks and Subplots: Harnessing Cause and Effect to Timing

The good news is, there's a nifty set of clear cause-and-effect rules that govern the seamless flitting back and forth between a flashback or sub-plot and the main storyline:

- The only reason to go into a flashback is that, without the information it provides, what happens next won't make sense. Thus there is a specific need—or cause—that triggers the flashback.

- This cause needs to be clear, so we know, from the second the flashback begins, why we're going into it. We must have a pretty good sense of why we need this information *now*. And as the flashback unfolds, we always need to sense how it relates to the story that's been put on hold.

- When the flashback ends, the information it provided must immediately—and necessarily—affect how we see the story

213

from that point on. The flashback needs to have given us information without which what's about to happen wouldn't have quite made sense. This isn't to say it can't also have given us information whose significance we won't learn until later—but it can't be only that.

Foreshadowing: A Genuine Get-Out-of-Jail-Free Card

It happens all the time. You've carefully plotted out your protagonist Stephanie's gauntlet of challenge, and she's doing quite well indeed, until suddenly she discovers that in order to uncover the *real* truth about Uncle Cedric, she has to hide in that teeny tiny broom closet under the stairs for who knows how long, which is fine, until you remember that you gave her claustrophobia back in chapter 2 to explain why she couldn't take her niece Becky on the ill-fated submarine ride at Disneyland. Now what? If you ignore the fact that she has claustrophobia and let her spend the evening crouched in that stuffy closet anyway, your readers instantly spot it, and this being the digital age, waste no time in shooting you a snide email telling you so. But if you go back and let her take Becky on the damn ride, it will change everything that's happened since. So what do you do? This is where a little foreshadowing comes in very handy.

At some point between the aborted submarine ride and when she has to head into the closet, Steph might reflect on her need to overcome her fear of small spaces. And so each time she trudges up the thirty flights of stairs to her office she thinks, *Geez, if only that elevator wasn't so small.* That way, when at last she has to squeeze in amid the brooms, mops, and dust rags, instead of its being a groaner, it's another of those well-established hurdles that we're rooting she'll make it over.

WAIT—I MUST HAVE MISSED SOMETHING

Don't underestimate the damage even a short-lived logic glitch can inflict. For instance, let's say we know that Rhonda loves Todd with all her heart. It's their anniversary, and she's on her way to the market to buy ingredients for the romantic dinner she's planning to make him, when she catches a glimpse of Todd kissing a mysterious stranger. But instead of fireworks, or waterworks, Rhonda doesn't give it a second thought. However, the reader does. Because suddenly Rhonda is acting completely out of character. We're dying to call the author and ask him what the hell is going on.

Were we to do just that, chances are he'd chuckle and tell us not to worry. Rhonda has a perfectly good (albeit currently unfathomable) reason for it, which he'd point out is right there in the very next paragraph, if only we'd had the patience to read another measly line or two.

So, who's right, the writer or the reader?

The reader, every time. Here's why: as far as the reader is concerned, the second Rhonda sees Todd smooching someone else and *doesn't* react, the story comes to a screeching halt. Suddenly it doesn't make sense, catapulting the reader out of the story and into her conscious mind. The result? She pauses, right then and there, and thinks about it. She wonders whether she's missed something along the way. Was Todd the one who had occasional bouts of amnesia, maybe? And although that pause may last only a nanosecond, it stops the story's momentum cold. That's why, even if the answer is in the very next sentence, it won't do a thing to remedy the problem. How could it? *Because it's already happened.*

Don't let it.

HOW TO MAKE THE READER BELIEVE YOUR PROTAGONIST CAN, IN FACT, FLY

The flip side is that there's absolutely nothing your protagonist can't do—be it fly, walk through walls, or recite the dictionary backward—provided you've foreshadowed this unusual talent long before it becomes the only way out of a sticky situation. So if you're going to shatter, bend, or reinterpret any of the laws of the universe—established laws we take for granted—you need to give us fair warning. This is especially crucial when you're dabbling in science fiction, fantasy, or magical realism. While you're utterly free to turn your characters loose in a world completely of your own creation, this gives you the added responsibility of not only establishing the rules of logic by which that world operates, but also of rigorously sticking to them. That way, when you foreshadow a change, the reader will have a good idea what it's a change *from*.

The same is true when you want your characters to deviate from the norm. We all know that if you don't eat, you get hungry; if you don't drink, you get thirsty; and if you don't look both ways before you cross the street, you might get mowed down by a twit texting a tweet. As a result, when your protagonist does not conform to our basic cause-and-effect expectations, we get grouchy. We don't have a choice; our brain uses its knowledge of how the world works as a default base to judge the credibility of characters.[12] We don't like having to ask (with apologies to Shakespeare), "Hey, when the protagonist gets mowed down, does he not *bleed*?"

This is not to say he has to bleed. Far from it. It means if he isn't going to bleed, you better have already given us a pretty good reason why. It doesn't cut it to tell us, as he lies there on the pavement, giggling: "Oh, by the way, John is really a Martian, and did I mention that they're made of rubber, so when that car hit him it didn't hurt; it tickled?"

Thus when characters are going to do something decidedly out of the ordinary, we need to already know one of two things:

1. They have the ability to do it, because we've seen them in action. For instance, we don't want to learn that Donna can walk through reinforced steel walls at the very moment Wendy locks her in an airtight basement and flips the switch to suck all the remaining oxygen out of the room. But if we've watched

Donna walk through a wall or two before, preferably when nothing big hung in the balance, then we're right there with her, breathing a sigh of relief as Wendy turns the lock, knowing that Donna has outsmarted her again.

2. If we haven't actually *seen* Donna walk through walls, we must have been given enough "tells" along the way that once she *does*, it's not only believable, but also satisfying. No wonder Donna was so good at hide-and-seek as a kid! And that time she leaned against the wall and almost fell through it? *Now* I get it! (Of course, the caveat is that back when those things happened, they must have made sense in and of themselves, rather than standing out like a sore thumb with a neon tattoo reading: "Don't worry, this will be explained later.")

Even so, it can *still* be very tempting to goad a character into doing things they'd never do in a million years because the plot needs them to. This is something you see in movies a whole lot. Edgar has a double PhD in rocket science and applied psychology from Stanford, speaks six languages, is a black belt, and writes best-selling mystery novels in his spare time, yet when a nervous stranger breathlessly asks him to take a crudely sealed package across the border from Mexico into El Paso, he agrees without a moment's hesitation. Because if Edgar isn't arrested crossing the border, well, the entire third act would go up in smoke.

Resist this temptation. Listen to your characters, who will implore you to give them a believable reason for everything they do, every reaction they have, every word they say, and every memory that suddenly pops into their head and changes how they see everything. This is exactly what, when done right, foreshadowing, flashbacks, and subplots do for your reader. Readers go willingly on these seeming side trips because they know from their own lives that the remembrance of things past often offers up nuggets of wisdom they can't afford to ignore.

CHAPTER 11: CHECKPOINT

Do all your subplots affect the protagonist, either externally or internally, as he struggles with the story question? Readers don't want subplots just because they're interesting or lyrical or provide a nice break from the intensity of the main action. Although they may be *all* these things, first and foremost, the reader expects they'll be there for a story reason. So ask yourself: Even if it's tangential, how does this subplot affect the protagonist's pursuit of his goal? What specific information does it give your reader that she needs to know in order to really grasp what's happening to the protagonist?

When you leap into a subplot or flashback, can the reader sense why it was necessary at that very moment? Make sure the logic is on the page, not just in your head. When you leave the main storyline, you want the reader to follow you willingly, not kicking and screaming.

When returning to the main storyline, will your reader see things with new eyes from that moment on? You want her to come back to the main storyline feeling as though she has new insight that gives her the inside track on what's going on.

When the protagonist does something out of character, has it been foreshadowed? Make sure you've given the reader solid tells along the way, so her reaction will be "Aha!" rather than "Give me a break!"

Have you given your reader enough information to understand what's happening, so that nothing a character does or says leaves her wondering whether she missed something? You never want your reader to have to pause, trying to figure out what she's missed, and then—god forbid—leaf back through the book to try to figure it out.

THE WRITER'S BRAIN ON STORY

COGNITIVE SECRET:

It takes long-term, conscious effort to hone a skill before the brain assigns it to the cognitive unconscious.

STORY SECRET:

There's no writing; there's only rewriting.

The first principle is that you must not fool
yourself, and you are the easiest person to fool.

—RICHARD P. FEYNMAN

WE'VE SPENT A LOT OF TIME looking at story through the reader's hardwired expectations. What about ours? How does the writer's brain fit into the equation when it comes to creating a story that leaps off the page? Turnabout being fair play, perhaps it's time we slid our own DNA under the microscope for a quick look-see. I'll go first.

I noticed a strange thing not too long ago. When I misspell a word, the more I try to figure out how to spell it, the more mangled it gets. If, instead, I simply retype it without thinking—fast, like pulling off a Band-Aid—nine times out of ten it's spelled correctly. For a while I went around telling people that my brain doesn't know how to spell but my fingers do. Turns out it's my brain, after all, which knows more than I think it does—provided I don't think about it.

As neuropsychiatrist Richard Restak says, "In many cases we decrease accuracy and efficiency by thinking too hard."[1] He points to an example many of us remember with chagrin: taking multiple-choice tests back in school and constantly second-guessing our answers. According to studies, if we'd just stuck with our original gut instinct and moved on to the next question like the instructions suggested, we would have gotten that A we knew we deserved, instead of—well, never mind. The one lesson we can still take away from that frustrating experience is that often the harder we try, the worse we do.[2]

So, does that mean winging it really is the best bet? Should you forget everything we've been talking about and join the ranks of the "pantsers"—that is, writers who write solely by the seat of their pants?

Probably not before you read the fine print: Restak goes on to say that following your gut works only if you've prepared for the test and know the material.[3]

That notion might have been what prompted seventeenth-century writer Thomas Fuller to observe that all things are difficult before they are easy. Indeed, Nobel laureate Herbert Simon estimates that it takes about ten years to really master a subject. By then we've gathered upward of fifty thousand "chunks" of knowledge, which the brain has deftly indexed so our cognitive unconscious can access each chunk on its own whenever necessary. Simon goes on to explain that this is "why experts can . . . respond to many situations 'intuitively'—that is, very rapidly, and often without being able to specify the process they have used to reach their answers. Intuition is no longer a mystery."[4]

Antonio Damasio agrees: "Outsourcing expertise to the unconscious space is what we do when we hone a skill so finely that we are no longer aware of the technical steps needed to be skillful. We develop skills in the clear light of consciousness, but then we let them go underground, into the roomy basement of our minds. . . ."[5]

It's through this process that story becomes intuitive for writers and that muscle memory is built. The good news is, the clock probably started ticking on your ten-year apprenticeship a long time ago. You likely already know—at least somewhere in your cognitive subconscious—that a huge part of the writing process is rewriting, with gusto, if possible.

That's why in this chapter we'll examine the deceptive thrill of finishing a first draft; discuss why seeking no-holds-barred criticism is crucial; explore why rewriting is an essential part of the writing process; consider the pros and cons of writers' groups; look at the benefits of professional literary consultants; and finally discover a painless way to toughen your hide before heartless strangers begin mercilessly attacking the very essence of your being (read: critiquing your work).

First, the High of Crossing the Finish Line

You've finished your first draft. You're elated. Giddy. You can't believe it's actually done. You wanted to give up a million times, but you didn't. You slogged from the terrifying emptiness of the blank page to the two most beautiful words in a writer's vocabulary: "The End." You do exactly what you should: go out and celebrate.

The next morning, basking in the afterglow of this genuinely magnificent accomplishment, you decide it might be a good idea to reread your manuscript before submitting it to literary agents, just in case there are typos. But within a page or two, you're facing the mystery of the ages: *how can scenes that seemed so insanely suspenseful when you wrote them, so completely engaging, so downright profound, suddenly sound so flat and banal?* Did monkeys get into your computer while you slept?

Before you hit the Delete key and decide to take up interpretive dance instead, you should know that this happens to everyone. It's important (not to mention reassuring) to keep in mind that writing is a process. It is rarely possible to address all of a story's trouble spots in a single draft, so don't be hard on yourself. It's not you; it's the nature of the beast. If there is one thing every successful writer's process includes, it's rewriting. Talent aside, in my experience, what separates writers who break through from those who don't is perseverance mixed with the wholehearted desire of a zealot to zero in on what isn't working and fix it.

Don't believe me? What about John Irving, who said, "Half my life is an act of revision."[6] Or Dorothy Parker, who said, "I can't write five words but that I change seven."[7]

Or Caroline Leavitt, author of *Girls in Trouble*, who rewrote her ninth novel several times before showing it to her agent, then rewrote it four more times based on the agent's notes. The book sold immediately. And then she wrote another four drafts, this time for her editor.

Or literary young adult author John H. Ritter, who estimates he rewrites each novel fifteen times before publication. Or UCLA screenwriting chairman Richard Walter, who reports that former student and exceedingly successful screenwriter David Koepp will happily rewrite for the studios until about the seventeenth draft, at which point he gets a little cranky.

To sum up the point these writers are making, let's turn to Ernest Hemingway, who, with characteristic blunt eloquence, so famously said, "All first drafts are shit." Which doesn't let you off the hook. It's not a license for unbridled self-expression, or not to try hard from word one because it doesn't really "count." It does, big time—because from here on out, it's the raw material you'll be working with, straying from, reshaping, paring, parsing, and then lovingly polishing. First drafts count, even if they're usually pretty bad. But remember, there's a huge difference between "trying hard" (which you want to do) and "trying to make it perfect from the first word on" (which is impossible and just might shut you down). The goal isn't beautiful writing; it's to come as close as you can to identifying the underlying story you're trying to tell.

So whether it's your first draft or your fifteenth, relax. Instead of thinking each draft has to be "it," just try to make your story a little bit better than it was in the previous draft. After all, stories are layered, and everything that happens affects everything else—and on every level, no less. That means when you remedy one problem, you'll most likely have shifted something somewhere else that will then need to be addressed, and so on. The point is, it's impossible to address every trouble spot in a single draft, so why make yourself crazy trying?

However, writers have a hardwired advantage when it comes to keeping track of who does what to whom and why. It may not be a super power, but it comes in pretty handy, especially as you begin your rewrite. Let me explain. . . .

The Writer's Brain Advantage

Recently, evolutionary psychologist Robin I. M. Dunbar asked himself the question we've been wrestling with from the beginning: considering that the ability to appreciate a story is universal, why are good writers so rare? His research reveals that one of the key factors revolves around something called "intentionality." This boils down to our ability to infer what someone else is thinking. In a pinch, most people can keep track of five states of mind at once. Says Dunbar, "When the audience ponders Shakespeare's Othello, for example, they are obliged to work at fourth order intentional levels: I (the audience) *believe* that Iago *intends* that Othello *supposes* that Desdemona *wants* [to love someone else]. When Shakespeare puts the play on stage before us, he will, in critical scenes, have four individuals interacting, thus obliging us to work at fifth order level—the very limits to which most of us can cope."[8]

What makes good writers different? We can hold in our minds what we know and what our characters believe and at the same time keep track of what our readers believe, sometimes to the sixth or seventh level. Sounds like a video game, doesn't it?

So, especially during a rewrite when you're digging deep, it helps to keep track of each character's version of reality.

Like a Circle in a Spiral, Like a Wheel within a Wheel

As the author, you know the big picture. You know what's *really* going on. You know where the treasure is buried and where, no matter how diligently the protagonist searches, she'll come up empty. You know who's lied to the hero and who's told the truth. You know which facts are true and which are not.

Your characters, on the other hand, often have no idea of what's *actually* going on, which means that they'll do things that presuppose an entirely different world than the one they're living in.

As writers, this is something we sometimes lose sight of. Because *we* know what the truth is, and what the future will be, we forget that our characters don't. And no wonder—considering that at any one time there may be four or five worlds in play.

What does that mean, exactly?

Well, there's the real world, meaning the objective world within the story. That's the actual, overarching world in which everything takes place, where things are as they are, sans interpretation or spin. Chances are *none* of your characters is completely familiar with this particular world. In fact, they couldn't be, since it is impossible to know absolutely everything about everything (even in a fictional world). Thus each character knows only a portion of what is "really" happening. What's more, some of what they "know" is probably very wrong—and this is often where the conflict comes from. On top of that, each character then puts her or his own personal spin on everything.

Of course, that doesn't stop the protagonist from acting on the assumption that what he believes is true actually *is* true, and often he pays a big price for it. For instance, Romeo—fully believing that Juliet is dead when he returns, heartbroken, to Verona—pursues the only option he sees as viable. He drinks a vial of poison and very dramatically dies. He has no way of knowing that in two itty bitty minutes the potion Juliet swallowed will wear off and she'll yawn and stretch. Then they could have hightailed it out of there and lived happily ever after. In this case, the "real world" and the world Romeo *thought* was real were, tragically, two very different places.

Reality Check

This brings us to a very helpful set of questions to ask yourself as you begin writing or rewriting each scene:

- What is actually going on in the story's "real world"—that is, objectively?

- What does each character *believe* is going on?

- Where are there contradictions? (Joe, believing that his brother Mark is their dad's favorite, is forever trying to win his dad's approval; Mark knows that their dad is really an evil alien, so he has been protecting Joe from him ever since he was born.)

- Given what each character believes is true (as opposed to what might actually be true), how would they act in the scene?

- Does what each character does in the scene make sense, given what he or she believes is true?

In fact, it's a good idea to make a chart for your entire story, called:

WHO KNOWS WHAT, WHEN?

First, make a timeline chronicling what actually happens in the "real world" during the span of the story. For instance, Romeo meets Juliet; they fall in love and secretly marry; she asks him to stop a fight between their houses; he tries and ends up killing her kinsman; her parents betroth her to a man who leaves her cold; Romeo, not knowing that Juliet has been betrothed, flees for the time being; Juliet, with the help of the friar, fakes her death to get out of marrying the other guy, sending a letter to Romeo explaining the plan; Romeo doesn't get the letter, rides back to Verona, finds the drugged Juliet in a crypt, and thinking she's dead, kills himself; Juliet wakes up, realizes what's happened, and does likewise. Their chastened families make up.

Beneath *your* overarching timeline, make a corresponding timeline for each major character, charting what they *believe* is true throughout the story. This will not only reveal exactly where and when characters are at cross purposes, but also help you make sure your characters' reactions are in accordance with what *they* believe is true in the moment.

Finally, there is one more person whose shifting beliefs you want to chart: the reader. Ask yourself, scene by scene: what does the reader believe is happening? This question is so important that you might even want to close the laptop, get out of your PJs, and head into the real world to test the waters. After all, you now know exactly what readers are hardwired to hunt for, so you can use them to do reconnaissance for you.

Starter Feedback—Priming the Pump

Before you begin asking for gut-wrenching critiques (anything short of "It's the best thing I ever read! Where can I buy a copy or, better yet, a case?"), there's an incredibly helpful type of feedback you can request at just about any stage without having to weather anyone's actual opinion. What's more, the info it yields tends to be clear, concise, and specific, and even your old Uncle Rolly can give it. Ideally, it's best to recruit friends and family who don't even know what your story is about. All you have to do is ask them to read what you have and at the end of each scene to jot down the answers to these questions:

- What do you think is going to happen next?

- Who do you think the important characters are?

- What do you think the characters want?

- What, if anything, leaps out as a setup?

- What information did you think was really important?

- What information were you dying to know?

- What did you find confusing? (This is as close to a real critique as we'll get.)

Their answers will be extremely helpful in figuring out how much of the story hasn't quite made it from your head onto the page. Not to mention turning up plot holes, logic gaps, redundancies, digressions, and long flat stretches that stop the story cold. But be sure to tell them this is *all* the feedback you want right now. If you give Uncle Rolly carte blanche, you might have to hear his theory on how much better it would be if it was set on the planet Zelon instead of in Cleveland, if the hero was an intergalactic warrior instead of a kindergarten teacher, and if lots of big things blew up instead of that one measly fight when Wally threw a handful of sand at Jane during recess.

Other People's Opinions

But at some point—on draft three or six or twenty-seven—you will need to let other people read your story for real. This is because, no matter how painstakingly objective you are, how ruthless when it comes to ferreting out digressions, how willing to subject everything in your story to heartless scrutiny, it's still, um, *you* doing it. And no matter how accomplished you are, the one thing you can't do is read your story as if you've never heard it before. It's already there in your mind, fully realized, before you start reading. Since you know what everything means and where it's *really* going, how can you possibly tell whether the words on the page are capable of conjuring the same thing in someone else's mind? Someone who has nothing *but* the words on the page to go on? Remember the Heath brothers' "Curse of Knowledge"? You can't tell your story's effect on a fresh reader, because you know way too much.

That's why you have to subject your story to the most merciless thing on earth: a reader's eyes. It could be those of a trusted writer friend, a writer's group, a paid professional, or even better, all three. This can feel a bit like asking the entire neighborhood to take pot shots at your children while they're playing in the yard all by themselves. And guess what? They will. Readers are more than willing to take a whack at our darlings, because to them, they aren't darlings at all. They're merely the things that get in the way of the story.

As humorist Franklin P. Jones famously said, "Honest criticism is hard to take, particularly from a relative, a friend, an acquaintance, or a stranger."

EMBRACING FEEDBACK

The importance of getting outside feedback—*and then actually listening to it*—can't be overstated. What's more, and trickier still, you want to be sure that the person giving you feedback is capable of it. This doesn't just mean they have the ability to zero in on what pulled them out of the story, but that when they see it go off the rails, they'll tell you.

Consider the story of a woman we'll call Zoe, who had written a memoir. She grew up in a small community where her mother was a local celebrity, thrusting Zoe into the limelight from kindergarten on. Even more compelling, her personal life sounded like a very successful movie of the week—the kind that makes you laugh, makes you cry, and leaves you with an authentic sense of hope. The trouble was, she did not know how to tell a story. Without a genuine narrative thread (read: no story question, no internal issue), the book didn't build. So it wasn't long before what little momentum it started with dissolved, leaving in its wake a series of disjointed vignettes. Somewhere around chapter 3 it went flat, and it stayed that way. It didn't matter that each individual scene was well written, because without an overarching context to give all the scenes meaning, the reader didn't know what to make of them or where the memoir was headed.

But Zoe did. She saw it very clearly. Why wouldn't she? She'd lived it. She'd shown the manuscript to close friends and an old college professor, all of whom told her how much they loved it and how well written it was. So when her agent gave her specific notes for the rewrite, instead of listening, she spent hours explaining why each suggested change was unnecessary and why everything that seemed to be missing was actually there. She felt it was good enough. She was a very likable young woman who had been through a hell of a lot (as her memoir attested). And it soon became clear she wasn't going to back down. So the manuscript was submitted, as is, to editors at twenty publishing houses. These editors didn't know her at all, nor had they heard her lengthy explanations for the things that they instantly saw weren't working. Every editor it was submitted to turned it down, each rejection letter echoing the notes the writer had already heard from her agent and blithely dismissed.

Sure, her friends thought it was perfect. But they were already familiar with her story, so they automatically filled in whatever blanks she'd inadvertently left. And, even more dangerous, they loved her. Which meant they were predisposed to like what she'd written, not to mention quite impressed that she'd sat down and written an entire book in the first place. In other words, what made it a page turner for them *wasn't* her storytelling skill.

Does that mean that when they told her they thought it was a great book they were lying? Of course not. It means the standard they applied to her manuscript wasn't the same one they use when they walk into a bookstore, pull a random book from the shelf, and start reading.

However, they didn't know that. And to further complicate the matter, chances are they couldn't have told her what their criteria for loving a book actually *are*, anyway. It's like that old saw, *I can't define pornography, but I know it when I see it.*[9] Which means it's a gut feeling. Or, in the case of pornography, sometimes that feeling is a little further south.

The truth is, it's almost impossible to differentiate between the gut feeling you get when you're reading a fabulous book and the feeling you get when you're reading a manuscript written by a close friend. It's surprisingly easy to misattribute the cause of a gut feeling. For instance, there's a classic experiment in which an attractive woman approached men in the middle of a scary, heart-pounding suspension bridge over a deep gorge, and after asking them to fill out a questionnaire—supposedly for a class project—she gave them her phone number. She then did likewise with an equal number of men after they had crossed the bridge and were sitting on a bench, recovering. Around 65 percent of the men on the bridge called her, compared with 30 percent of those on the bench, whose hearts were no longer pounding when she approached them.[10] That is to say, a majority of them had mistaken an adrenaline rush of fear for the giddiness of attraction. In the same way, friends and family tend to misattribute the adrenaline surge they feel when they read your book to their appreciation of your prowess as a writer rather than the thrill of knowing you actually wrote it. This isn't to say you may not actually have written a crackerjack book, but chances are, they won't be able to tell the difference.

In other words, love is blind.

And when it's not, it tends to be supportive. When you read a friend's writing, your first allegiance is to your friend. So even when your gut tells you that it's probably not time for her to quit her day job, you take into consideration how hard she worked, how much the book means to her, and the fact that you don't want to hurt her feelings. Or start a fight. The same is true with acquaintances. No one wants to be the bearer of bad news; it inherently stirs up strong emotions—in this case, most likely the kind of conflict-induced tension that the manuscript in question probably isn't generating. But as we know, whereas in books, conflict is what draws us in, in real life, it's something most people will go out of their way to avoid. Which is why when you read a friend's manuscript and find it completely devoid of tension, the last thing you want to do is actually create some by mentioning it.

So you find nice things to say: *Loved the premise. Fabulous thesis. Great sense of place; I really felt like I was in downtown Barstow. And Tiffany's clever retort when she caught Tad rifling through her underwear drawer—priceless!* Your friend beams, and you haven't told a single lie. Except by omission. But hey, you tell yourself, you're not a professional critic. Maybe the book really *is* great, but you're just too much of a dolt to see it. And so you breathe a heartfelt sigh of relief and enthusiastically give the manuscript the benefit of the doubt.

But as a writer, is that something *you* would really want? The benefit of the doubt? Hey, why not! When you've sweated blood over something, given it your all, you want to hear that it's great. Perfect. Brilliant, in fact. Then again, would you want your doctor to have been given the benefit of the doubt throughout medical school? Or the pilot of the jumbo jet you're about to board?

But wait—doesn't your story belong to you? Who says writers have to please everyone? First and foremost, don't we have to write for ourselves, to speak our truth? Maybe. But ask yourself, when you read a novel, do you really ever want to know the writer's truth? Do you even think about it? The truth we're looking for is something we can relate to *ourselves*. Writers who focus on "their truth" tend to forget that as far as the reader is concerned, writing is about communication, not self-expression. That brings us to another myth whose neck we might want to wring:

MYTH: Writers Are Rebels Who Were Born to Break the Rules

REALITY: Successful Writers Follow the Damn Rules

Writers are often rebels. We buck the tide by trade. We have a fresh take on the familiar, and our goal is to translate that vision into a story so others can step into our world. Since we're all about originality, why should we have to follow a tired old set of standards, anyway? Can't we just peel that girdle off and breathe freely? After all, we make up stories; can't we make up the rules, too?

It's at about this point in the argument that someone always starts talking about Cormac McCarthy. *He* doesn't follow the rules, and he won the Pulitzer. My response is always, *He does follow the rules, but he's done it in such an idiosyncratic way that it's easy to take his style for a new set of rules.* Yes, there are masters out there with such utterly distinct voices that they have the ability to instill an intoxicating sense of urgency in ways that *seem* to defy analysis. It's in their DNA, which is why it cannot be duplicated. They're in a rarefied minority. If we could write like them, we'd have long since been published, and universities would offer graduate seminars on our work.

On the other hand, the vast majority of extremely successful writers *don't* write like them, either.

And here's something a little more sobering. For every successful writer who *seems* to flout the rules, there are millions along the way who tried to *actually* flout them, and whose manuscripts crashed and burned as a result. You just never heard about them because, well, they crashed and burned. Chances are they either ignored the feedback they got or, worse, never asked for it.

HOW CAN YOU IMPROVE IF YOU DON'T KNOW WHAT'S WRONG?

Writers need impartial feedback, and one of the logical places to get it is in a writers' group. The members of an effective writing group need to be astute and able to not only point out what isn't working but also tell you why. The rub, of course, is that they also have to be right. The places where something isn't working are not hard to spot. What's hard is explaining exactly *why* it isn't working. This often leads to misguided advice, which results in the writer either making the problem worse or simply substituting one thing that isn't working for another. So when you join a writers' group—especially if you don't know anyone in it yet—your best bet is to sit back and listen. You will learn far more about them by how they critique each other's work than how they critique yours. Why?

First of all, because you can actually hear it. Being singled out in a group, especially for the first time, can be overwhelming. Remember what we said about the mortification of discovering you've made a mistake in public? That's what a critique can feel like. Everyone is looking at you, and your face goes red, there's a loud buzzing in your ears, and suddenly the room gets very hot. People are talking, but you can't make out the words. It's hard enough to hear, let alone be objective.

On the other hand, when they're critiquing someone else, it's infinitely easier to judge whether their comments are on target or flying wide of the mark. You'll have your own opinion of the work you hear and so be able to gauge whether their comments are insightful, astute, and expressed in a way that is supportive while at the same time, pulling no punches.

Keep in mind, too, that a writers' group, by definition, will hear your work in pieces. Thus it can be difficult for them to tell whether the story is building, if the setups are paying off, or if that beautifully written passage about Jamie's first kiss that had them all crying has *anything* to do with the story of how she and her sixty-eight-year-old grandmother climbed Mt. Everest.

HIRE A PRO

The other option when it comes to getting feedback is a trend that is gaining momentum. A colleague at a literary agency in New York recently told me, "More than ever it is important for writers to hone their craft and submit only their most polished professional draft. Do not count on anyone—agent or even [in-house] editor—to 'fix' it. Everyone is so tight for time that material has to be rewritten several times, and edited, before anyone in the business sees it to consider.

Using freelance editors and consultants to help get a manuscript in shape is increasingly common."

The good news is, there are many extremely capable freelance literary consultants out there who can provide objective, professional feedback that can help you not only rewrite your story but also improve your writing skills in the process. The bad news is, you'll find a gazillion to choose from—some great, some not—just by typing "literary consultant" into Google. My advice is to make sure the person you hire has a background in publishing—either as an agent or as an editor. If you're a screenwriter, look for someone with genuine development experience. If you're considering hiring a story analyst, find out what production company or studio they read for, and how long. Experience matters. Because while any intern can (and does) decide whether or not a script or novel works, when it doesn't, very few can tell you exactly why—and fewer still, what to do about it.

Better Them Than Us, For Now

One way to toughen your hide before you venture into this territory is to start reading reviews—book reviews, movie reviews, reviews of all sorts. Why? For perspective. Think of it as a training course. Imagine you are the author of the book that's being taken to task. Because, let me tell you, reviewers are merciless—as they should be. Often gleefully so.

For instance, in his review of the movie version of *The Da Vinci Code*, A. O. Scott of the *New York Times* manages to take a pretty good swing at both author Dan Brown and screenwriter Akiva Goldsman. First calling Brown's bestseller a "primer on how not to write an English sentence," he goes on to chide Goldsman for penning "some pretty ripe dialogue all on his own."[11]

Ouch. But at least that's just about the prose rather than the authors themselves. For that, here is *Slate's* Dana Stevens on the movie version of Elizabeth Wurtzel's bestselling memoir *Prozac Nation*:

> Granted, *Prozac Nation* is an extremely silly movie, but let's face it: self-dramatizing middle-class girls who stay up for days on end writing *Harvard Crimson* articles about Lou Reed ("I feel his cold embrace, his sly caress") are inherently silly people. . . . And whenever the film takes Wurtzel's tragic posing seriously, it flounders.[12]

Double ouch. In one shot, Stevens slams the book, the movie, and Wurtzel herself. In print. For everyone to see. And given that the Internet is now home to just about everything anyone says about anything, both reviews will be at the world's fingertips, a mere couple of keystrokes away, 24/7, forever.

Be prepared: regardless of how successful you get, people are going to be analyzing your work, for better or worse, from here on out. Some will come at it with bizarre, idiosyncratic potshots; others will zero in with dead-on accuracy and illuminate massive trouble spots you won't believe you could have missed.

If you have trouble now hearing it from a friend, in private, imagine how it'll feel from a stranger, in public. Thus your goal is to toughen up. That's not to say you won't feel gut punched at first. There's no real way around it. Miguel de Cervantes Saavedra had this warning for his fellow writers: "No fathers or mothers think their own children ugly; and this self-deceit is yet stronger with respect to the offspring of the mind."[13]

It's Always Darkest Before the Sunshine

Is it worth it to rewrite an entire novel or screenplay two, three, or four times? What about five or six? Just how many times are we talking about? It's impossible to say. So perhaps an anecdote will suffice—one that highlights just how long the road can be and how sweet the reward at the end.

Back in 1999, Michael Arndt felt he'd paid his dues, having spent ten years in the movie business as a script reader and assistant. So, having accumulated a small nest egg in the process, he quit his job and hunkered down to write a screenplay. He wrote six stories and ditched each one. The seventh—which he wrote in three days—he had a good feeling about.[14] So he kept at it. For over a *hundred* drafts. His motto was *No point in doing something if you're not going to do it right.* And he was determined to get it right.[15]

Which is probably why, six years after he began writing it, he won the Oscar for best original screenplay for *Little Miss Sunshine.* Why? Because his allegiance wasn't to himself, or to his first draft, or even to his ninety-ninth. It was to the story itself. And to us. A world full of strangers who he knew would never, ever give him the benefit of the doubt. So his story didn't ask us to. All it required of us was that we sit back, relax, and give it our undivided attention.

With that kind of care and determination, imagine how far *your* story can go. You don't need to be a genius, although you may well be one. What you do need is perseverance. The one thing that makes a person a writer is *writing.* Butt in chair. Every day. No excuses. Ever. As Jack London famously said, "Don't loaf and invite inspiration; light out after it with a club."[16] Hemingway concurred: "Work every day. No matter what has happened the day or night before, get up and bite on the nail."[17]

It's only then that the real story you're telling slowly emerges. Here's a secret: when you've tapped into what it is we're wired to respond to in a story, what we're hungry for from the very first sentence, it *is* your truth we hear. As neuroscientist David Eagleman says, "When you put together large numbers of pieces and parts, the whole can become something larger than the sum. . . . The concept of emergent properties means that something new can be introduced that is not inherent in any of the parts."[18]

What emerges is your vision, seen through the eyes of your readers, *experienced* by your readers. So what are you waiting for? Write! Although they may not know it yet, your public is eager to find out what happens next.

- End -

Endnotes

INTRODUCTION

1. M. Gazzaniga, *Human: The Science Behind What Makes Your Brain Unique* (New York: Harper Perennial, 2008), 220.

2. J. Tooby and L. Cosmides, 2001. "Does Beauty Build Adapted Minds? Toward an Evolutionary Theory of Aesthetics, Fiction and the Arts," *SubStance* 30, no. 1 (2001): 6–27.

3. Ibid.

4. S. Pinker, *How the Mind Works* (New York: W. W. Norton, 1997/2009), 539.

5. M. Djikic, K. Oatley, S. Zoeterman, and J. B. Peterson, "On Being Moved by Art: How Reading Fiction Transforms the Self," *Creativity Research Journal* 21, no. 1 (2009): 24–29.

6. Common quote based on J. L. Borges, "Tlön, Uqbar, Orbis Tertius," in *Ficciones*, trans. Emecé Editores (New York: Grove Press, 1962), 22.

7. PhysOrg.com, "Readers Build Vivid Mental Simulations of Narrative Situations, Brain Scans Suggest," January 6, 2009, http://www.physorg.com/print152210728.html.

CHAPTER 1: HOW TO HOOK THE READER

1. T. D. Wilson, *Strangers to Ourselves: Discovering the Adaptive Unconscious* (Cambridge, MA: Belknap Press of Harvard University Press, 2002), 24.

2. R. Restak, *The Naked Brain: How the Emerging Neurosociety Is Changing How We Live, Work, and Love* (New York: Three Rivers Press, 2006), 24.

3. D. Eagleman, *Incognito: The Secret Lives of the Brain* (New York: Pantheon, 2011), 132.

4. A. Damasio, *Self Comes to Mind: Constructing the Conscious Brain* (New York: Pantheon, 2010), 293.

5. Ibid., 173.

6. Ibid., 296.

7. Pinker, *How the Mind Works*, 543 (see introduction, n. 4).

8. B. Boyd, *On the Origin of Stories: Evolution, Cognition, and Fiction* (Cambridge, MA: Belknap Press of Harvard University Press, 2009), 393.

9. J. Lehrer, *How We Decide* (Boston and New York: Houghton Mifflin Harcourt, 2009), 38.

10. R. Montague, *Your Brain Is (Almost) Perfect: How We Make Decisions* (New York: Plume, 2007), 111.

11. C. Leavitt, *Girls in Trouble* (New York: St. Martin's Griffin, 2005), 1.

12. J. Irving, "Getting Started," in *Writers on Writing*, ed. R. Pack and J. Parini (Hanover, NH: University Press of New England, 1991), 101.

13. Restak, *Naked Brain*, 77.

14. D. Devine, "Author's Attack on Da Vinci Code Best-Seller Brown," WalesOnline.co.uk, September 16, 2009, http://www.walesonline .co.uk/news/wales-news/2009/09/16/author-s-astonishing-attack-on-da-vinci-code-best-seller-brown-91466-24700451.

CHAPTER 2: HOW TO ZERO IN ON YOUR POINT

1. M. Lindstrom, *Buyology: Truth and Lies About Why We Buy* (New York: Broadway Books, 2010), 199.

2. P. Simpson, *Stylistics*. London: Routledge, 2004), 115.

3. Boyd, *On the Origin of Stories*, 134 (see ch. 1, n. 8).

4. Wilson, *Strangers to Ourselves*, 28 (see ch. 1, n. 1).

5. Lehrer, *How We Decide*, 37 (see ch. 1, n. 9).

6. Boyd, *On the Origin of Stories*, 134.

7. Damasio, *Self Comes to Mind*, 168 (see chap. 1, n. 4).

8. R. Maxwell and R. Dickman, *The Elements of Persuasion: Use Storytelling to Pitch Better, Sell Faster & Win More Business* (New York: HarperBusiness, 2007), 4.

9. Pinker, *How the Mind Works*, 539 (see introduction, n. 4).

10. E. Strout, *Olive Kitteridge* (New York: Random House, 2008), 281.

11. Ibid., 224.

12. E. Waugh, *The Letters of Evelyn Waugh*, ed. by M. Amory (London: Phoenix, 1995), 574.

13. M. Mitchell, *Gone with the Wind* (New York: Simon & Schuster Pocketbooks, 2008), 1453.

14. W. Golding, *Lord of the Flies* (New York: Perigee Trade 2003), 304.

15. "Gabriel (Jose) García Márquez," *Contemporary Authors Online, Gale, 2007. Reproduced in Biography Resource Center.* (Farmington Hills, MI: Gale, 2007), http://www.gale.cengage.com/free_resources/chh/bio/marquez_g.htm.

16. Mitchell, *Gone with the Wind*, 1453.

CHAPTER 3: I'LL FEEL WHAT HE'S FEELING

1. Lehrer, *How We Decide*, 13 (see ch. 1, n. 9).

2. A. Damasio, *Descartes' Error: Emotion, Reason, and the Human Brain* (New York: Penguin, 1994), 34–50.

3. Pinker, *How the Mind Works*, 373 (see introduction, n. 4).

4. Gazzaniga, *Human*, 226 (see introduction, n. 1).

5. Damasio, *Self Comes to Mind*, 254 (see ch. 1, n. 4).

6. Gazzaniga, *Human*, 179.

7. Wilson, *Strangers to Ourselves*, 38 (see ch. 1, n. 1).

8. E. George, *Careless in Red* (New York: Harper, 2008), 94.

9. A. Shreve, *The Pilot's Wife* (New York: Little, Brown & Company, 1999), 1.

10. E. Leonard, *Freaky Deaky* (New York: William Morrow Paperbacks, 2005), 117.

11. George, *Careless in Red*, 99.

12. Restak, *Naked Brain*, 65 (see ch. 1, n. 2).

13. Pinker, *How the Mind Works*, 421.

14. J. W. Goethe, "The Poet's Year," in *Half-Hours with the Best Authors*, vol. IV, ed. C. Knight (New York: John Wiley, 1853), 355.

15. Gazzaniga, *Human*, 190.

16. C. Heath and D. Heath, *Made to Stick: Why Some Ideas Survive and Others Die* (New York: Random House, 2007), 20.

17. Common quotation based on M. Twain, *Following the Equator* (Hartford, CT: American Publishing Company, 1898), 156.

18. W. Grimes, "Donald Windham, 89, New York Memoirist (Obituary)," *New York Times*, June 4, 2010.

19. J. Franzen, Life and Letters, "Mr. Difficult," *New Yorker*, September 30, 2002, 100.

CHAPTER 4: WHAT DOES YOUR PROTAGONIST *REALLY* WANT?

1. Pinker, *How the Mind Works*, 188 (see introduction, n. 4).

2. M. Iacoboni, *Mirroring People: The New Science of How We Connect with Others* (New York: Farrar, Straus & Giroux, 2008), 34.

3. Gazzaniga, *Human*, 179 (see introduction, n. 1).

4. Boyd, *On the Origin of Stories*, 143 (see ch. 1, n. 8).

5. PhysOrg.com, "Readers Build Vivid Mental Simulations" (see introduction, n. 7).

6. Pinker, *How the Mind Works*, 61.

7. *Public Papers of the Presidents of the United States, Dwight D. Eisenhower*, 1957 (Washington, DC: National Archives and Records Service, Federal Register Division, 1958).

8. J. Barnes, *Flaubert's Parrot* (New York: Vintage, 1990), 168.

9. K. Oatley, "A Feeling for Fiction," *Greater Good,* The Greater Good Science Center, University of California, Berkeley, Fall/Winter 2005–6, http://greatergood.berkeley.edu/article/item/a_feeling_for_fiction.

10. M. Proust, *Remembrance of Things Past*, trans. C. K. Scott-Montcrieff (New York: Random House, 1934), 559.

11. J. Nash, *The Threadbare Heart* (New York: Berkley Trade, 2010), 9.

CHAPTER 5: DIGGING UP YOUR PROTAGONIST'S INNER ISSUE

1. Wilson, *Strangers to Ourselves*, 31 (see ch. 1, n. 1).

2. Gazzaniga, *Human*, 272 (see introduction, n. 1).

3. K. Schulz, *Being Wrong: Adventures in the Margin of Error* (New York: ecco, 2010), 109.

4. Damasio, *Self Comes to Mind*, 211 (see ch. 1, n. 4).

5. T. S. Eliot, *Four Quartets* (Boston: Mariner Books, 1968), 59.

6. B. Forward, "Beast Wars, Part 1," *Transformers: Beast Wars*, season 1, episode 1, directed by I. Pearson, aired September 16, 1996.

7. G. Plimpton, "Interview with Robert Frost," in *Writers at Work: The Paris Review Interviews*, 2nd series (New York: Viking, 1965), 32.

8. T. Brick, "Keep the Pots Boiling: Robert B. Parker Spills the Beans on Spenser," *Bostonia*, Spring 2005.

9. K. A. Porter, interview by B. T. Davis, *The Paris Review* 29 (Winter-Spring 1963).

10. J. K. Rowling, interview by Diane Rehm, *The Diane Rehm Show*, WAMU Radio Washington, DC, transcript by Jimmi Thøgersen, October 20, 1999, http://www.accio-quote.org/articles/1999/1299-wamu-rehm.htm.

11. J. K. Rowling, interview by C. Lydon, *The Connection* (WBUR Radio), transcript courtesy *The Hogwarts Library*, October 12, 1999, http://www.accio-quote.org/articles/1999/1099-connectiontransc2.htm; J. K. Rowling, interview, Scholastic, transcript, February 3, 2000, http://www.scholastic.com/teachers/article/interview-j-k-rowling.

12. Gazzaniga, *Human*, 190.

13. Ibid., 274.

CHAPTER 6: THE STORY IS IN THE SPECIFICS

1. Pinker, *How the Mind Works* 285 (see introduction, n. 4).

2. Ibid., 290.

3. Gazzaniga, *Human*, 286 (see introduction, n. 1).

4. Damasio, *Self Comes to Mind*, 188 (see ch. 1. n. 4).

5. V. S. Ramachandran, *The Tell-Tale Brain: A Neuroscientist's Quest for What Makes Us Human* (New York: W.W. Norton, 2011), 242.

6. Damasio, *Self Comes to Mind*, 121.

7. Ibid., 46–47.

8. G. Lakoff, "Metaphor, Morality, and Politics, Or, Why Conservatives Have Left Liberals In The Dust," *Social Research* 62, no. 2 (Summer 1995): 177–214.

9. Pinker, *How the Mind Works*, 353.

10. J. Geary, "Metaphorically Speaking," TEDGlobal 2009, July 2009, transcript and video, http://www.ted.com/talks/lang/eng/james_geary_metaphorically_speaking.html.

11. Aristotle. *Poetics* (Witch Books, 2011), 53.

12. E. Brown, *The Weird Sisters* (New York: Amy Einhorn Books/ Putnam, 2011), 71.

13. NPR, "Tony Bennett's Art of Intimacy," September 16, 2011, http:// www.npr.org/2011/10/29/141798505/tony-bennetts-art-of-intimacy.

14. Heath and Heath, *Made to Stick*, 139 (see ch. 3, n. 16).

15. E. Leonard, *Elmore Leonard's 10 Rules of Writing* (New York: William Morrow, 2007), 61.

16. Pinker, *How the Mind Works*, 377.

17. G. G. Marquez, *Love in the Time of Cholera* (New York: Vintage Books, 2007), 6.

CHAPTER 7: COURTING CONFLICT, THE AGENT OF CHANGE

1. Damasio, *Self Comes to Mind*, 292 (see ch. 1, n. 4).

2. Lehrer, *How We Decide*, 210 (see ch. 1, n. 9).

3. Wilson, *Strangers to Ourselves*, 155 (see ch. 1, n. 1).

4. B. Patoine, "Desperately Seeking Sensation: Fear, Reward, and the Human Need for Novelty," The Dana Foundation, http://www .dana.org/media/detail.aspx?id=23620.

5. Restak, *The Naked Brain*, 216 (see ch. 1, n. 2).

6. E. Kross et al., "Social Rejection Shares Somatosensory Representations with Physical Pain," Proceedings of the National Academy of Sciences of the United States of America 108, no. 15 (April 12, 2011): 6270–6275. http://www.ncbi.nlm.nih.gov/pmc/articles/ PMC3076808.

7. J. Mercer, "Ac-cent-tchu-ate the Positive (Mister In-Between)," by J. Mercer and H. Arlen, October 4, 1944, *Over the Rainbow*, Capitol Records.

8. Damasio, *Self Comes to Mind*, 54.

9. Gazzaniga, *Human*, 188–89 (see introduction, n. 1).

10. D. Rock and J. Schwartz, "The Neuroscience of Leadership with David Rock and Jeffrey Schwartz," *Strategy + Business*, webinar, November 2, 2006, http://www.strategy-business.com/webinars/ webinar/webinar-neuro_lead?gko=37c54.

CHAPTER 8: CAUSE AND EFFECT

1. J. P. Wright, *The Skeptical Realism of David Hume* (Manchester: Manchester University Press, 1983), 209.

2. Damasio, *Self Comes to Mind*, 133 (see ch. 1, n. 4).

3. Gazzaniga, *Human*, 262 (see introduction, n. 1).

4. K. Schulz, "On Being Wrong," TED2011, March 2011, transcript and video, http://www.ted.com/talks/kathryn_schulz_on_being_ wrong.html.

5. Damasio, *Self Comes to Mind*, 173.

6. Boyd, *On the Origin of Stories*, 89 (see ch. 1, n. 8).

7. L. Neary, "Jennifer Egan Does Avant-Garde Fiction—Old School," NPR, Morning Edition, July 6, 2010, http://www.npr.org/templates/ story/story.php?storyId=128702628.

8. A. Chrisafis, "Overlong, Overrated, and Unmoving: Roddy Doyle's Verdict on James Joyce's Ulysses," *The Guardian*, February 10, 2004, http://www.guardian.co.uk/uk/2004/feb/10/booksnews.ireland.

9. J. Franzen, "Q. & A. Having Difficulty with Difficulty," *New Yorker* Online Only, September 30, 2002.

10. A. S. Byatt, "Narrate or Die," *New York Times Magazine*, April 18, 1999, 105–107.

11. Neary, "Jennifer Egan."

12. For the original translation of this phrase from Chekhov's letter to his brother, see W. H. Bruford, *Anton Chekhov* (New Haven, CT: Bowes and Bowes, 1957), 26.

13. Boyd, *On the Origin of Stories*, 91.

14. Damasio, *Self Comes to Mind*, 211.

15. The Isaiah Berlin Literary Trust, "Anton Chekhov," The Isaiah Berlin Virtual Library, 2011, quoted from S. Shchukin, *Memoirs*, 1911, http://berlin.wolf.ox.ac.uk/lists/quotations/quotations_by_ib.html.

16. D. Gilbert, "He Who Cast the First Stone Probably Didn't," *New York Times*, July 24, 2006, The Opinion Pages.

17. M. Twain, *The Complete Letters of Mark Twain* (Teddington, UK: Echo Library, 2007), 415.

18. J. Boswell, *The Life of Samuel Johnson* (New York: Oxford University Press USA, 1998), 528.

CHAPTER 9: WHAT CAN GO WRONG, MUST GO WRONG—AND THEN SOME

1. Restak, *The Naked Brain*, 216 (see ch. 1, n. 2).

2. R. I. M. Dunbar, "Why Are Good Writers So Rare? An Evolutionary Perspective on Literature," *Journal of Cultural and Evolutionary Psychology*, 3, no. 1 (2005): 7–21.

3. Gazzaniga, *Human*, 220 (see introduction, n. 1).

4. Pinker, *How the Mind Works*, 541 (see introduction, n. 4).

5. R. A. Mar et al., "The Function of Fiction Is the Abstraction and Simulation of Social Experience," *Perspectives on Psychological Science* 3, no. 3 (2008): 173–192.

6. P. Sturges, *Five Screenplays by Preston Sturges* (Berkeley, CA: University of California Press, 1986), 541.

7. Schulz, *Being Wrong*, 26 (see ch. 8, n. 4).

8. H. Vendler, *Dickinson: Selected Poems and Commentaries* (Cambridge: Belknap Press of Harvard University Press, 2010), 54.

9. Eagleman, *Incognito*, 145 (see ch. 1, n. 3).

10. J. W. Pennebaker, "Traumatic Experience and Psychosomatic Disease: Exploring the Roles of Behavioural Inhibition, Obsession, and Confiding," *Canadian Psychology/Psychologie canadienne* 26, no. 2 (1985): 82–95.

11. Damasio, *Self Comes to Mind*, 121 (see ch. 1, n. 4).

12. Pinker, *How the Mind Works*, 540.

13. P. McGilligan, *Backstory: Interviews with Screenwriters of Hollywood's Golden Age* (Berkeley and Los Angeles: University of California Press, 1986), 238.

14. T. Carlyle, *The Best Known Works of Thomas Carlyle: Including Sartor Resartus, Heroes and Hero Worship and Characteristics* (Rockville, MD: Wildside Press, 2010), 122.

15. Plutarch, *Plutarch's Lives, Volume 3* (Cambridge, MA: Harvard University Press, 1967), 399.

16. C. G. Jung, *Alchemical Studies (Collected Works of C. G. Jung*, vol. 13) (Princeton, NJ: Princeton University Press, 1983), 278.

CHAPTER 10: THE ROAD FROM SETUP TO PAYOFF

1. Boyd, *On the Origin of Stories*, 89 (see ch. 1, n. 8).

2. S. J. Gould, *Bully for Brontosaurus: Reflections in Natural History* (New York: W. W. Norton, 1991), 268.

3. Damasio, *Self Comes to Mind*, 64 (see ch. 1, n. 4).

4. Gazzaniga, *Human*, 226 (see introduction, n. 1).

5. Heath and Heath, *Made to Stick*, 286 (see ch. 3, n. 16).

6. D. Rock and J. Schwartz, "The Neuroscience of Leadership with David Rock and Jeffrey Schwartz," *Strategy + Business*, webinar, November 2, 2006, http://www.strategy-business.com/webinars/webinar/webinar-neuro_lead?gko=37c54.

7. R. Chandler, *Raymond Chandler Speaking* (Berkeley and Los Angeles, CA: University of California Press, 1997), 65.

8. A. Gorlick, "Media Multitaskers Pay Mental Price, Stanford Study Shows," *Stanford* Report, August 24, 2009, http://news.stanford.edu/news/2009/august24/multitask-research-study-082409.

9. Boyd, *On the Origin of Stories*, 90.

10. J. Stuart and S. E. de Souza, *Die Hard*, directed by J. McTiernan. Silver Pictures and Gordon Company, 20th Century Fox, 1988.

11. C. Leavitt, *Girls in Trouble* (New York: St. Martin's Griffin, 2005), 98.

CHAPTER 11: MEANWHILE, BACK AT THE RANCH

1. S. B. Klein et al., "Decisions and the Evolution of Memory: Multiple Systems, Multiple Functions," *University of California, Santa Barbara Psychological Review* 109, no. 2 (2002): 306–329.

2. Damasio, *Self Comes to Mind*, 211 (see ch. 1, n. 4).

3. Gazzaniga, *Human*, 187–88 (see introduction, n. 1).

4. Ibid., 224.

5. Pinker, *How the Mind Works*, 540 (see introduction, n. 4).

6. D. Chase, R. Green, and M. Burgess, "All Due Respect," *The Sopranos*, season 5, episode 13, directed by J. Patterson, aired June 6, 2004 (HBO, Chase Films, and Brad Grey Television).

7. Lehrer, *How We Decide*, 237 (see ch. 1, n. 9).

8. N. Bransford, "Setting the Pace," March 5, 2007, http://blog.nathanbransford.com/2007/03/setting-pace.html.

9. Boyd, *On the Origin of Stories*, 90 (see ch. 1, n. 8).

10. G. Lucas, G. Katz, and W. Huyck, *American Graffiti*, directed by G. Lucas. American Zoetrope and LucasFilm, Universal Pictures, 1973.

11. W. Mosley, *Fear Itself* (New York: Little, Brown & Company, 2003), 140.

12. Gazzaniga, *Human*, 190.

CHAPTER 12: THE WRITER'S BRAIN ON STORY

1. Restak, *Naked Brain*, 23 (see ch. 1, n. 2).

2. P. C. Fletcher et al., "On the Benefits of Not Trying: Brain Activity and Connectivity Reflecting the Interactions of Explicit and Implicit Sequence Learning," *Cerebral Cortex* 15, no. 7 (2005): 1002–1015.

3. Restak, *Naked Brain*, 23.

4. H. A. Simon, *Models of Bounded Rationality, Vol 3: Empirically Grounded Economic Reason* (Cambridge, MA: MIT Press, 1997), 178.

5. Damasio, *Self Comes to Mind*, 275 (see ch. 1, n. 4).

6. J. Irving, *Trying to Save Piggy Sneed* (New York: Arcade Publishing, 1996), 5.

7. S. Silverstein, *Not Much Fun: The Lost Poems of Dorothy Parker* (New York: Scribner, 2009), 47.

8. Dunbar, Why Are Good Writers So Rare? (see ch. 9, n. 2).

9. Based on a concurring opinion by Justice P. Stewart, Jacobellis v. Ohio, 378 U.S. 184 (1964).

10. D. G. Dutton et al., "Some Evidence for Heightened Sexual Attraction under Conditions of High Anxiety," *Journal of Personality and Social Psychology* 30, no. 4 (1974): 510–517.

11. A. O. Scott, "'Da Vinci Code' Enters Yawning," *New York Times*, May 17, 2006, http://www.nytimes.com/2006/05/17/arts/17iht-review.1767919.html?scp=7&sq=goldsman%20da%20vinci%20brown&st=cse.

12. D. Stevens, *Slate*, March 22, 2005, http://www.slate.com/articles/news_and_politics/surfergirl/2005/03/what_have_you_done_with_my_office.single.html#pagebreak_anchor_2.

13. M. Cervantes Saavedra, *The Life And Exploits of the Ingenious Gentleman Don Quixote De La Mancha*, vol. 2 (Charleston, SC: Nabu Press, 2011), 104.

14. Wikipedia, s.v. "Michael Arndt," accessed October 25, 2011, http://en.wikipedia.org/wiki/Michael_Arndt.

15. A. Thompson, "'Closet screenwriter' Arndt Comes into Light," *Hollywood Reporter*, November 17, 2006.

16. J. London, "Getting into Print," *The Editor*, March 1903.

17. B. Strickland, ed., *On Being a Writer* (Cincinnati, OH: Writers Digest Books, 1992).

18. Eagleman, *Incognito* (see ch. 1, n. 3).

Acknowledgments

If there's one thing I've learned, first from story and then from neuroscience, it's that every decision we make is based on everything that's happened to us up to that moment. So it's no surprise that this book owes its existence to many, many people who have graciously given me their encouragement, expertise, and support.

For one thing, I wouldn't know nearly as much about story if not for a gifted group of friends, family, and colleagues: Jeannie Luciano, Paul F. Abrams, Mona Friedman, Judy Toby, Bill Contardi, Pamela Katz, Richard Walter, Amy Bedik, Sara Cron, Judy Nelson, Edith Barshov, Martha Thomas, LaDonna Mabry, Abra Bigham, Brett Hudson, Doug Michael, Vicky Choy, Iris Chayet, Marnie McLean, Angela Rinaldi, Frances Phipps, Mark Poucher, A. Karno, and Newman Wolf.

I'm grateful to Linda Venis, mastermind of the phenomenal UCLA Extension Writers' Program, and her fabulous staff: Mae Respicio, Kathryn Flaherty, and Sara Bond to name a few. Teaching in the Writers' Program allowed me to expand and refine my ideas, thanks to spot-on feedback from the most inspiring and talented students anywhere. Thanks especially to Michele Montgomery, who turned to me one night after class and said, "You're always telling everyone they have time to write a book; why aren't *you* writing one?" And to students Tommy Hawkins, Jill Beyer, and Sheel Kamal Seidler, whose wry, spirited questions always kept me on track.

I owe a special debt of gratitude to those who read the manuscript over the course of its many iterations and generously took the time to give me much-needed feedback. It has been improved immensely thanks to Lynda Weinman, Caroline Leavitt, Lisa Doctor, Rachel Kann, Colin Kindley, Carlyn Robertson, Michelle Fiordaliso, Charlie Peters, Randy Lavender, Jon Keeyes, Cherilyn Parsons, Dr. Ronald Doctor, Murray Nosel, Chris Nelson, Wendy Taylor, Robert Rotstein, Karen Karl, Robert Wolff, and Leigh Leveen.

Story has long been my profession, but I've come to neuroscience more recently. I was more than humbled that Michael Gazzaniga, cofounder of the field of cognitive neuroscience, took the time to read the manuscript and pronounce it fit. Thank you.

To the indefatigable, insightful, beautifully brutally honest freelance editor Jennie Nash: what would I have done without you? Undying thanks to my daughter Annie, who cheerfully read the manuscript countless times, always with the uncanny knack of finding obvious-when-pointed-out-but-otherwise-invisible logic glitches that no one else noticed, especially me. Many of my ideas were sharpened during conversations with my son Peter, whose love of story is as keen as mine. There is no one I'd rather talk story with, and no one I learn more from. Thanks to writers Jason Benlevi, for always believing in me, especially when I didn't, and Thomas Koloniar, for invaluable lessons in grit, perseverance, and loyalty.

Words don't do justice to how grateful I am to my supportive, savvy, brilliant agent, Laurie Abkemeier of DeFiore & Company. Somehow, she magically made the whole process stress-free. How often can you say that about anything? This book would be vastly different (read: not nearly as complete) if not for the shrewd wisdom of my whip-smart editor, Lisa Westmoreland. Thanks to her and the crackerjack team at Ten Speed Press, this book is infinitely better than it would otherwise have been.

Deepest heartfelt gratitude to my husband, Stuart Demar, who lovingly cooked every meal and did all the housework on top of his busy schedule, so I could keep writing into the wee hours. Only a true tough guy would do that. And finally, everlasting thanks to my lifelong best friend, Don Halpern, who makes everything possible. Damon never had a truer Pythias.

About the Author

A graduate of UC Berkeley, **LISA CRON** spent a decade in publishing—first at W. W. Norton in New York, then at John Muir Publications in Santa Fe, New Mexico—before turning to television, where among other things she's been supervising producer on shows for Court TV and Showtime. She's been a story consultant for Warner Brothers and the William Morris Agency in New York City and  for Village Roadshow, Icon, the Don Buchwald Agency, and others in Los Angeles. Lisa is featured in the book *Ask the Pros: Screenwriting* (Lone Eagle, 2004). Her personal essays have appeared on Freshyarn.com and the *Huffington Post*, and she has performed them at the 78th Street Playhouse in NYC, and in Los Angeles at Sit 'n Spin, Spark!, Word-A-Rama, Word Nerd, and Melt in Your Mouth (a monthly personal essay series she coproduced). For years she's worked one-on-one with writers, producers, and agents developing book and movie projects. Lisa has also been a literary agent at the Angela Rinaldi Literary Agency. She is currently an instructor in the UCLA Extension Writers' Program. She lives in Santa Monica, California, with her husband, two scruffy but well-loved cats, and a mischievous dog. Visit wiredforstory.com.

Index